6th International Workshop on Health Text Mining and Information Analysis (LOUHI 2015)

Held at EMNLP 2015

Lisbon, Portugal
17 September 2015

ISBN: 978-1-5108-1191-1

Printed from e-media with permission by:

Curran Associates, Inc.
57 Morehouse Lane
Red Hook, NY 12571

Some format issues inherent in the e-media version may also appear in this print version.

Printed by Curran Associates, Inc. (2015)

For permission requests, please contact the Association for Computational Linguistics at the address below.

Association for Computational Linguistics
209 N. Eighth Street
Stroudsburg, Pennsylvania 18360

Phone: 1-570-476-8006
Fax: 1-570-476-0860

acl@aclweb.org

Additional copies of this publication are available from:

Curran Associates, Inc.
57 Morehouse Lane
Red Hook, NY 12571 USA
Phone: 845-758-0400
Fax: 845-758-2634
Email: curran@proceedings.com
Web: www.proceedings.com

EMNLP 2015

Sixth International Workshop on Health Text Mining and Information Analysis (Louhi)

Proceedings of the workshop

17 September 2015
Lisbon, Portugal

Preface

The Sixth International Workshop on Health Text Mining and Information Analysis provides an interdisciplinary forum for researchers interested in automated processing of health documents. Health documents encompass electronic health records, clinical guidelines, spontaneous reports for pharmacovigilance, biomedical literature, health forums/blogs or any other type of health related documents. The Louhi workshop series fosters interactions between the Computational Linguistics, Medical Informatics and Artificial Intelligence communities. It started in 2008 in Turku, Finland and has been organized five times: Louhi 2010 was co-located with NAACL in Los Angeles, CA; Louhi 2011 was co-located with Artificial Intelligence in Medicine (AIME) in Bled, Slovenia; Louhi 2013 was held in Sydney, Australia during NICTA Techfest; and Louhi 2014 was co-located with EACL in Gothenburg, Sweden.

The aim of the Louhi 2015 workshop is to bring together research work on topics related to text mining of health documents, particularly emphasizing multidisciplinary aspects of health documentation and the interplay between nursing and medical sciences, information systems, computational linguistics and computer science. The topics include, but are not limited to, the following Natural Language Processing techniques and related areas:

- Techniques supporting information extraction, e.g. named entity recognition, negation and uncertainty detection
- Classification and text mining applications (e.g. diagnostic classifications such as ICD-10 and nursing intensity scores) and problems (e.g. handling of unbalanced data sets)
- Text representation, including dealing with data sparsity and dimensionality issues
- Domain adaptation, e.g. adaptation of standard NLP tools (incl. tokenizers, PoS-taggers, etc) to the medical domain
- Information fusion, i.e. integrating data from various sources, e.g. structured and narrative documentation
- Unsupervised methods, including distributional semantics
- Evaluation, gold/reference standard construction and annotation
- Syntactic, semantic and pragmatic analysis of health documents
- Anonymization / de-identification of health records and ethics
- Supporting the development of medical terminologies and ontologies
- Individualization of content, consumer health vocabularies, summarization and simplification of text

- NLP for supporting documentation and decision making practices
- Predictive modeling of adverse events,e.g. adverse drug events and hospital acquired infections

The call for papers encouraged authors to submit papers describing substantial and completed work but also focus on a contribution, a negative result, a software package or work in progress. We also encouraged to report work on low-resourced languages, addressing the challenges of data sparsity and language characteristic diversity.

We received 39 submissions, an unprecedented high number for the LOUHI series. Each submission went through a double-blind review process which involved three program committee members. Based on comments and rankings supplied by the reviewers, we accepted 19 papers (11 long papers and 8 short papers). The overall acceptance rate is 49% and the acceptance rate for long papers is 50%. During the workshop, 8 papers have been presented orally, and 11 papers have been presented as posters.

Finally, we would like to thank the members of the program committee for the quality of theirs reviews in a very short period. We are very grateful to Marie-Francine Moens for accepting to give an invited talk. We would also like to thank the authors for their submissions and the quality of their work.

Cyril Grouin, Thierry Hamon, Aurélie Névéol, Pierre Zweigenbaum

Organizers:

Cyril Grouin, LIMSI-CNRS, Orsay, France
Thierry Hamon, LIMSI-CNRS, Orsay, France & Université Paris 13
Aurélie Névéol, LIMSI-CNRS, Orsay, France
Pierre Zweigenbaum, LIMSI-CNRS, Orsay, France

Program Committee:

Sophia Ananiadou, University of Manchester, U.K.
Sabine Bergler, Concordia University, Canada
Thomas Brox Røst, Norwegian University of Science and Technology, Norway
Kevin B. Cohen, University of Colorado/School of Medicine, USA
Francisco Couto, University of Lisbon, Portugal
Hercules Dalianis, Stockholm University, Sweden
Louise Deléger, INRA, France
Gaël Dias, Normandie University, France
Martin Duneld/Hassel, Stockholm University, Sweden
Richárd Farkas, Institute of Informatics, Hungary
Filip Ginter, University of Turku, Finland
Natalia Grabar, CNRS UMR 8163, STL Université de Lille3, France
Gintaré Grigonyté, Stockholm University, Sweden
Aron Henriksson, Stockholm University, Sweden
Rezarta Islamaj, NIH/NLM/NCBI, USA
Antonion Jimeno Yepes, IBM Research, Australia
Jussi Karlgren, KTH, Royal Institute of Technology, Sweden
Dimitrios Kokkinakis, University of Gothenburg, Sweden
Maria Kvist, Stockholm University, Sweden
Alberto Lavelli, Fondazione Bruno Kessler, Italy
David Martínez, University of Melbourne and MedWhat.com, Australia
Beáta Megyesi, Uppsala University, Sweden
Marie-Jean Meurs, UQAM & Concordia University, QC, Canada
Fleur Mougin, Université de Bordeaux, ERIAS, Centre INSERM U897, ISPED, France
Danielle L Mowery, University of Utah, USA
Henning Müller, University of Applied Sciences Western Switzerland, Switzerland
Mariana Neves, Hasso-Plattner-Institute at the University of Potsdam, Germany
Jong C. Park, KAIST Computer Science, Korea
Jon D. Patrick, Health Language Laboratories, Australia
Sampo Pyysalo, University of Turku, Finland
Stefan Schulz, Graz General Hospital and University Clinics, Austria
Tapio Salakoski, University of Turku, Finland
Sanna Salanterä, University of Turku, Finland

Isabel Segura-Bedmar, Universidad Carlos III de Madrid, Spain
Maria Skeppstedt, Gavagai and Linnaeus University, Sweden
Hanna Suominen, NICTA, Australia
Suzanne Tamang, Stanford University School of Medicine, USA
Özlem Uzuner, MIT, U.S.A.
Sumithra Velupillai, Stockholm University, Sweden
Karin Verspoor, University of Melbourne, Australia
Mats Wirén, Stockholm University, Stockholm, Sweden

Invited Speaker:

Marie-Francine Moens, Department of computer Science, Katholieke Universiteit Leuven

Table of Contents

Program of the workshop

Thursday, September 17, 2015

09:00–10:30 Session I - Corpus creation

In-depth annotation for patient level liver cancer staging
Wen-wai Yim, Sharon Kwan and Meliha Yetisgen

Predicting Continued Participation in Online Health Forums
Farig Sadeque, Thamar Solorio, Ted Pedersen, Prasha Shrestha and Steven Bethard

Redundancy in French Electronic Health Records: A preliminary study
Eva D'hondt, Xavier Tannier and Aurélie Névéol

Is it possible to recover personal health information from an automatically de-identified corpus of French EHRs?
Cyril Grouin, Nicolas Griffon and Aurélie Névéol

11:00–12:30 Session II - Poster

An Analysis of Biomedical Tokenization: Problems and Strategies
Noa P. Cruz Diaz and Manuel Maña López

Annotation of Clinically Important Follow-up Recommendations in Radiology Reports
Meliha Yetisgen, Prescott Klassen, Lucas McCarthy, Elena Pellicer, Tom Payne and Martin Gunn

On the Impact of Twitter-based Health Campaigns: A Cross-Country Analysis of Movember
Nugroho Dwi Prasetyo, Claudia Hauff, Dong Nguyen, Tijs van den Broek and Djoerd Hiemstra

Exploring Word Embedding for Drug Name Recognition
Isabel Segura-Bedmar, Víctor Suárez-Paniagua and Paloma Martínez

Creating a rule based system for text mining of Norwegian breast cancer pathology reports
Rebecka Weegar and Hercules Dalianis

Parser Adaptation to the Biomedical Domain without Re-Training
Jeff Mitchell and Mark Steedman

Expanding a dictionary of marker words for uncertainty and negation using distributional semantics
Alyaa Alfalahi, Maria Skeppstedt, Rickard Ahlbom, Roza Baskalayci, Aron Henriksson, Lars Asker, Carita Paradis and Andreas Kerren

Held-out versus Gold Standard: Comparison of Evaluation Strategies for Distantly Supervised Relation Extraction from Medline abstracts
Roland Roller and Mark Stevenson

Checking a structured pathology report for completeness of content using terminological knowledge
Sebastian Busse

Effectively Crowdsourcing Radiology Report Annotations
Anne Cocos, Aaron Masino, Ting Qian, Ellie Pavlick and Chris Callison-Burch

Identifying Key Concepts from EHR Notes Using Domain Adaptation
Jiaping Zheng and Hong Yu

12:30–14:00 ***Lunch break***

14:00–15:30 **Session III - Invited talk**

Information Extraction from Biomedical Texts: Learning Models with Limited Supervision
Marie-Francine Moens

16:00–17:30 **Session IV - Corpus processing**

Adverse Drug Event classification of health records using dictionary based preprocessing and machine learning
Stefanie Friedrich and Hercules Dalianis

NLP–Based Readability Assessment of Health–Related Texts: a Case Study on Italian Informed Consent Forms
Giulia Venturi, Tommaso Bellandi, Felice Dell'Orletta and Simonetta Montemagni

Mining and Ranking Biomedical Synonym Candidates from Wikipedia
Abhyuday Jagannatha, Jinying Chen and Hong Yu

Representing Clinical Notes for Adverse Drug Event Detection
Aron Henriksson

In-depth annotation for patient level liver cancer staging

Wen-wai Yim
Biomedical Informatics and Medical Education
University of Washington
wyim@uw.edu

Sharon Kwan
Radiology
University of Washington
shakwan@uw.edu

Meliha Yetisgen
Biomedical Informatics and Medical Education
University of Washington
melihay@uw.edu

Abstract

Cancer stages, which summarizes extent of cancer progression, is an important tool for evidence-based medical research. However, they are not always recorded in the electronic medical record. In this paper, we describe work for annotating a medical text corpus with the goal of predicting patient level liver cancer staging in hepatocellular carcinoma (HCC) patients.

Our annotation consisted of identifying 11 parameters, used to calculate liver cancer staging, at the text span level as well as at the patient level. Also at the patient level, we annotated stages for three commonly-used liver cancer staging schemes. Our inter-rater agreement showed text annotation consistency 0.73 F1 for partial text match and 0.91 F1 at the patient level.

After annotation, we performed several document classification experiments for the text span annotations using standard machine learning classifiers, including decision trees, maximum entropy, naive Bayes and support vector machines. Thereby, we identified baseline performances for our task at 0.63 F1 as well as strategies for future improvement.

1 Introduction

Despite their importance in research, cancer stages are not always recorded in the electronic medical record (EMR) in structured or unstructured format (Evans et al., 1998). Even when collected they are often inaccurate (Yau et al., 2002)(Sexton et al., 2006). On the other hand, review of patient notes for medical conditions is both time-consuming and expensive. One strategy to minimize these costs is to leverage natural language processing (NLP) to automate the process.

In this paper, we describe work for annotating a corpus with the goal of predicting patient level liver cancer staging in hepatocellular carcinoma (HCC) patients. Our group took a detailed annotation approach, which included text span level and patient level annotation of parameters used in staging, as well as patient level annotation of stages for three liver cancer staging schemes. In our results we present our inter-rater agreements and our analysis from studying our domain experts' annotations. Finally, in our last section we deliver preliminary information extraction baselines using several standard machine learning classifiers.

Clinically relevancy for this task is especially well exemplified by HCC for which has many competing treatment options but no universally accepted clinical guidelines (Han et al., 2011). Moreover, HCC progresses differently across various age groups, ethnicities, lifestyles, and associated co-mordidities (McGlynn and London, 2011). Automatic staging may facilitate evidence-based research for targeted disease management by leveraging the EMR for best outcomes. Its scaleable nature would allow the process to be adapted for volumes of historical data, efficiently unlocking more information than comparable prospective trial studies.

2 Background

Cancer staging is used to summarize the extent of disease for cancer patients. Each cancer domain may have different criteria for its stages. For example, ovarian cancer stages differentiates between whether one ovary is invaded, both, or the entire pelvic region (American Cancer Society, 2014).

For liver cancers, in addition to tumor morphology and spread, patient performance status as well as liver function variables are incorporated into

Proceedings of the Sixth International Workshop on Health Text Mining and Information Analysis (Louhi), pages 1–11,
Lisbon, Portugal, 17 September 2015.

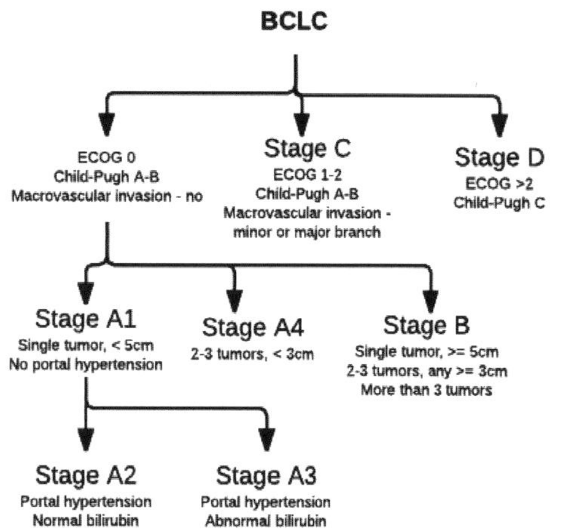

Figure 1: BCLC staging logic

various staging schemes. However, because there are various measures of tumor growth, liver failure, and overall patient well-being, over six different international liver cancer staging schemes exist (Sirivatanauksorn and Tovikkai, 2011). For our project we focus on capturing the parameters and classifications for three commonly used staging schemes: the American Joint Committee on Cancer (AJCC), the Barcelona Clinic Liver Cancer (BCLC), and the Cancer of the Liver Italian Program (CLIP) staging schemes (França et al., 2004). Figure 1 shows an example of the stage parameters, e.g. ***ECOG***, and the decision logic for classifying BCLC stages, e.g. ***Stage A1***.

In all, there are a total of 11 text parameters and 4 structured data laboratory parameters among the 3 staging schemes. Because Child-Pugh, one of the text parameters, is itself a classification system for severity of liver disease, when necessary, it must be calculated according to Table 1 logic.

The purpose of annotating stage parameters, in addition to overall stages, is two-fold. Firstly, more detailed annotation can presumably help with performance. Secondly, stage parameters may be used in more than one staging scheme, or may be re-used if a staging classification algorithm changes, given some little additional annotation.

Variable	Points		
	1	2	3
Albumin (g/dL)	> 3.5	2.8-3.5	< 2.8
Ascites	None	Mild/Moderate	Severe
Bilirubin (mg/dL)	< 2	2-3	> 3
Hepatic Encephalopathy	None	Grade 1-2	Grade 3-4
Prothrombin INR	< 1.7	1.7-2.3	> 2.3

Table 1: Child-Pugh parameters. Adding up the points for all variables, stage is assigned where Child-Pugh A: 5-6 points, Child-Pugh B: 7-9 points, and Child-Pugh C: 10-15 points.

3 Related work

We describe previous work by grouping systems by those that predict a cancer stage and those that extract cancer characteristics which make up stage parameters.

3.1 Cancer stage prediction

Previous work in automatic cancer staging from clinical documents focused on TNM cancer stage classification using document classification. A brief explanation: T, N, and M represent tumor size, lymph node spread, and metastasis, respectively. Each parameter takes different values depending on spread. For example T0 means no tumor, while T1-T4 are increasingly larger sizes. An example TNM stage for a patient is ***T2 N1 M0***.

Nguyen et al. (2007) predicted patient TNM stage by using multi-class document classification of concatenated records with support vector machines (SVM). They tested various hierarchical set-ups, i.e. binary for each variable vs. all versus all, etc, achieving accuracies of 64% and 82% for T and N sub-stages. The same group, McCowan et al. (2007) divided the document classification problem into a number of sentence-level classifications, in which a sentence is first classified for a particular parameter, e.g. T2, N1, etc. After predicting a value for each sentence, using SVM or some rules, the final stages were determined by post-processing heuristics. This strategy improved T and N accuracy to 74% and 87%. In their latest work, Nguyen et al. (2010) used a symbolic logic approach. Rules leveraged concept-normalization, negation, and normalization through the SNOMED-CT hierarchy. Their accuracy using these methods improved to 72%, 78%, and 94% for T, N, and M, respectively. Martinez and Li (2011) classified report level TN and ACPS stages, testing a mixture of document classification, sentence-level extraction, and rule-based methods and arrived at best F1 scores of 82%, 81%, and 75% for T, N, and ACPS staging, respectively.

Viewed from a larger scope, patient cancer stage prediction may be framed as a special case of clinical phenotype identification, which similarly involves distilling a patient's multiple clinical data sources, free-text and structured, to identify a specific disease or set of conditions. We will not discuss this with regards to our work here, but shall point the reader to an excellent review on phenotype cohorts using EMRs (Shivade et al., 2014).

3.2 Cancer information extraction

Two previous works focused specifically on liver cancer information extraction. One was a 2013 rule-based system (Ping et al., 2013) that extracted elements of liver cancer diagnosis, tumor characteristics, staging (BCLC and Child-Pugh), co-morbidities, and treatments using regular expression and rules. They captured concepts and relations in a diverse set of report types, with performances ranging 92-99% F1.

The other study was a 2014 hybrid system, Wang et al. (2014), in which HCC information was extracted from operation notes. First, relevant sentences of interest for a parameter were identified with keyword look-ups, then information was structured using a conditional random field algorithm. They achieved a 64% F1 performance.

A plethora of general cancer information extraction systems exist, concentrating on parameters such as tumor size, number and metastasis. Many use dictionary-based methods (Coden, 2009)(Ashish et al., 2014) for extracting entities before structuring them using specific algorithms. However statistical named entity recognition methods (Ou and Patrick, 2014) and document classification methods are also used (Jouhet et al., 2012)(Kavuluru et al., 2013).

Our annotation approach combines previous methods. Similar to McCowan et al. (2007) we annotated for stage parameters at a sub-document level before making an overall staging classification. However, we additionally annotated liver cancer specific information and marked at a text span level, as in Ping et al. (2013) and Wang et al. (2014). Unlike previous information extraction approaches, we annotated stage parameters at a patient level in addition to text span levels. Unlike previous cancer stage prediction systems, we classify over various report types instead of only histology and pathology reports.

4 Corpus Creation

4.1 Data and Processing

A cohort was drawn from new patients visiting the University of Washington (UW) Medical Center primary liver cancer clinic from 1/2011-12/2013 with approval by the UW Human Subjects Committee of Institutional Review Board. Included data for each patient comprised of: (1) all clinical notes from the day of visit to the clinic, including surgery, admit notes etc, (2) all laboratory results within 30 days prior to and following the visit day, and (3) radiology reports within 3 months prior to and 1 month following the visit day.

Patient records were manually reviewed by our clinical expert to exclude patients who had been seen prior to the start of the study and patients who had an obviously irrelevant diagnoses. Previously seen patients were excluded because they most likely had already started treatment, and our population of interest were patients at first presentation. Irrelevant report types were removed from the annotation set. For our study, we focused on the subset of patients that had at least one clinical report, at least one radiology report, and the full set of labs needed for staging. The resulting dataset included 236 patients and their associated 422 clinical and 309 radiology reports.

4.2 Guideline Creation

Guidelines for liver cancer stage and stage parameter annotations were developed primarily by an interventional radiologist with input from another interventional radiologist and a group of NLP scientists. Stage parameter values were discretized according to stage guidelines. The stage and stage parameters are described following.

Stage:

AJCC has classifications (I, II, IIIA, IIIB, IIIC, IVA, IVB) and is based on the TNM stage framework that primarily addresses tumor characteristics and spread but not liver functioning statuses.

BCLC has classifications (A1, A2, A3, A4, B, C, D) and is the only staging scheme that takes into account overall performance status (ECOG).

CLIP has classifications (0, 1, 2, 3, 4, 5, 6) and is the only staging scheme that takes into account the relative size of the tumor to the liver.

Stage parameters:

Ascites : accumulation of fluid in the peritoneal cavity (e.g., "no significant ascites," "does not endorse abdominal swelling") with values (None, Mild, Moderate-Severe)

Child-Pugh : a measurement of liver cirrhosis (e.g., "Child's B," "his CTP score would be 5") with values (A, B, C)

**ECOG (Eastern Cooperative Oncology

Group) Performance Status : a scaled measure of general well-being where 0 is fully active and 5 is dead (e.g., "ECOG 0," "She notes good energy") with values (0, 1, 2, $\geq$3)

Extrahepatic invasion : direct spread of cancer outside of the liver (e.g., "No evidence of extrahepatic extension," "the tumor may [...] extend from the liver to the right ribs or muscular wall") with values (No, Yes)

Hepatic encephalopathy : confusion or altered consciousness due to liver failure (e.g., "patient denies confusion, forgetfulness, or other symptoms of hepatic encephalopathy," "lactulose" in the medication list) with values (None, Mild, Severe)

Macrovascular invasion : spread of cancer to nearby blood vessels (e.g. "vascular invasion: possible involvement of middle hepatic vein branches," "no evidence of portal vein thrombosis") with values (No, Yes-minor_branch, Yes-major_branch)

Metastasis : spread of cancer to outside the liver, such as to lymph nodes (e.g., "lymph nodes suspicious for metastatic involvement: none," "no lymphadenopathy") with values (No, Yes-regional, Yes-distal)

Portal hypertension : elevation of hepatic venous pressure gradient to greater than 5 mm Hg (e.g., "no evidence of cirrhosis or portal hypertension," "patient had an EGD which showed small varices") with values (No, Yes)

Tumor morphology : number and size of tumor relative to the liver (e.g., "small segment 7 hepatic mass") with values (Massive $\geq$ 50% of liver, Multinodular $<$ 50% of liver, Uninodular $<$ 50% of liver)

Tumor number : number of liver tumors (e.g., "two new liver lesions noted [...] suggesting hepatomas") with values (Single, 2-3, >3)

Tumor size : diameter size of liver tumor (e.g., "1 lesion measuring 2.1 x 1.7 cm [...] HCC") with values (<3 cm, 3-5 cm, >5 cm).

Specifications on which sections to look for stage parameters in a report were formalized into annotation rules. Hepatic encephalopathy, Child-Pugh, and ECOG parameters were marked in clinical notes. Ascites was marked in both clinical notes and radiology notes. The remaining parameters were marked in radiology notes and only in clinical notes if they could not be found in radiology notes. For clinical notes, the annotators started at the "History of Present Illness" section before marking the rest of the note. For radiology notes, the annotators started at the "Impression" section, and if information could not be found there they would move on to "Findings" section or the rest of the report. Repeats of the same information were not annotated. The exception was for ECOG in which all descriptive mentions were also annotated. If multiple pieces of information contribute to the overall value, they were all marked.

4.3 Annotation workflow and software

Annotation occurred in two phases, carried out by two interventional radiologist. In the first phase, relevant parts of reports for the 11 stage parameters were identified by single annotation, where the corpus was divided evenly among the annotators by patient. Annotators marked text annotations using Brat (Stenetorp et al., 2012), a web-based graphical annotation tool, and assigned each a label, e.g. ***ECOG***, and a value, e.g. ***0***. Irrelevant patients, e.g. patients with irrelevant diagnosis, and files, e.g. addenda, abbreviated notes, and post-treatment radiology notes, were flagged for exclusion. Figure 2 shows example mark-ups.

During the second phase, the 3 overall stages and the 11 liver cancer parameters were annotated at the patient level by the consensus annotation of the two annotators. This stage required simultaneous review of all clinical and radiology notes as well as laboratory information related to the patient. The patient level annotation for the 11 liver cancer parameters was necessary to resolve missing and conflicting values from the phase I text annotations. Annotators used a specially built in-house python Tkinter (Hughes, 2000) interface, shown in Figure 3. Annotators had access to the full marked reports as well as a summarized version of their annotations displayed in the interface, along with pertinent laboratory values.

4.4 Text annotations inter-rater agreement

A subset of 20 patients were double annotated for phase I text annotations to calculate inter-rater agreement. After one round of annotations, the annotators met to resolve conflicts and fine-tuned annotation guidelines. We used precision, ($P = \frac{TP}{TP+FP}$), recall ($R = \frac{TP}{TP+FN}$), and F1-measure, ($F1 = \frac{2PR}{P+R}$), to measure inter-rater agreement, where TP is true positives, FP is false positives, and FN is false negatives. True positive matches were measured by label, value, and partial text

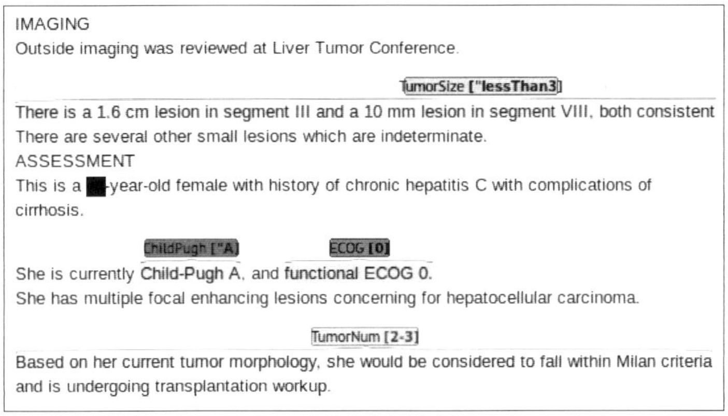

Figure 2: Example phase I text annotations using Brat

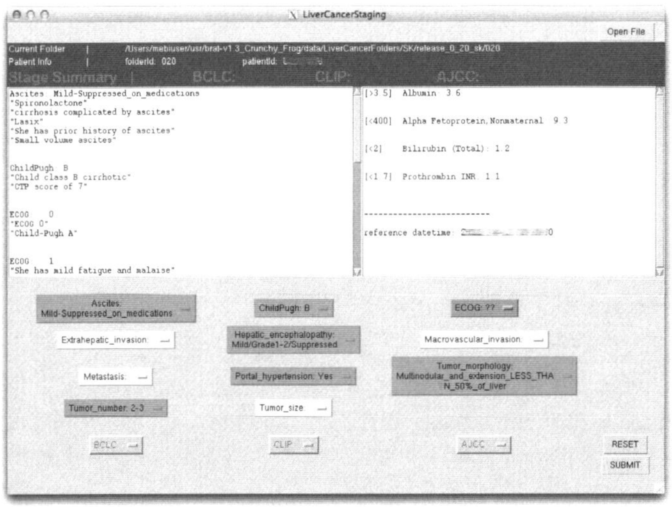

Figure 3: Example phase II patient level annotations. Display panels shows summarized text annotations (left) and lab values (right). Bottom buttons allow annotation of patient level label-values, including the 11 stage parameters (first four rows) and overall 3 stagings (bottom row).

span overlap. For example, two text annotations were considered matching if its label, e.g. ***ascites***, its value, e.g. ***none***, matches exactly and its text spans (document character offsets) overlap. It is also possible, to calculate the agreement of text annotations (still phase I annotations) resolved to the patient level, e.g. if two annotators both identify text spans in any patient 0 file with ***ascites - none***, it is a match.

After the annotator meeting, the microscore agreement for phase I text annotations improved from 0.45 to 0.73 F1. Table 2 shows the final inter-rater microscore agreement consolidated by label. Phase I patient level agreement improved from 0.76 to 0.91 F1. The final patient level agreement breakdown is shown in Table 3. Of the 20 sample patients, 3 patients were excluded due to irrelevant diagnosis.

Discrepancy between the text span and patient levels quantify how often the two annotators find the same information in separate files or different parts of the same document. The higher performance at the patient level was expected given the lower amount of precision needed for patient level agreement.

Ascites, ECOG, and hepatic encephalopathy, had lower agreements because they were often repeated in different expression formats in different report sections. Additionally, one annotator marked ascites drugs while the other did not. Extrahepatic invasion differences were due to one annotator identifying more descriptive information.

Label	TP	FP	FN	P	R	F1
Ascites	10	9	12	0.53	0.45	0.49
ChildPugh	7	0	0	1.00	1.00	1.00
ECOG	23	6	9	0.79	0.72	0.75
Extrahepatic_invasion	6	4	0	0.60	1.00	0.75
Hepatic_encephalopathy	12	3	5	0.80	0.71	0.75
Macrovascular_invasion	16	6	0	0.73	1.00	0.84
Metastasis	10	2	1	0.83	0.91	0.87
Portal_hypertension	11	7	5	0.61	0.69	0.65
Tumor_morphology	15	8	8	0.65	0.65	0.65
Tumor_number	17	6	7	0.74	0.71	0.72
Tumor_size	18	5	5	0.78	0.78	0.78
ALL	145	56	52	0.72	0.74	0.73

Table 2: Phase I inter-rater partial match of label-value per text span, consolidated by label

Label	TP	FP	FN	P	R	F1
Ascites	9	4	2	0.69	0.82	0.75
ChildPugh	6	0	0	1.00	1.00	1.00
ECOG	14	1	3	0.93	0.82	0.88
Extrahepatic_invasion	5	4	0	0.56	1.00	0.71
Hepatic_encephalopathy	8	2	2	0.80	0.80	0.80
Macrovascular_invasion	13	2	0	0.87	1.00	0.93
Metastasis	9	1	0	0.90	1.00	0.95
Portal_hypertension	11	3	0	0.79	1.00	0.88
Tumor_morphology	17	1	0	0.94	1.00	0.97
Tumor_number	17	0	0	1.00	1.00	1.00
Tumor_size	17	0	0	1.00	1.00	1.00
ALL	126	18	7	0.88	0.95	0.91

Table 3: Phase I inter-rater exact match of label-value per patient, consolidated by label

4.5 Phase I annotation statistics

A total of 36 patients and 91 documents were marked for exclusion during phase I of annotation. The total number of patients and associated documents left were 200 and 545, respectively. Of 545 documents, 303 were clinical notes and 242 were radiology notes. There was a total of 2108 text annotations. A breakdown is shown in Table 4.

4.6 Phase II annotation

At the time this paper was written, phase II patient level annotations were still under way, however the corresponding 20 patients used for inter-rater agreement had been staged. For this sample, we found cases where discrepancies in data sources or missing information led to indeterminable stage labels. This occurred for 2 out of 17 non-excluded patients in the 20 patient sample, in which BCLC staging could not be determined due to irreconcilable ECOG values.

5 Analysis of text annotation evidence

In this section, we describe the characteristics of text annotation evidence from the completed phase I of annotations, with the goal of highlighting idiosyncracies or potential challenges for building an information extraction system.

Label	Value	Freq
Ascites	Mild-Suppressed	56
	Moderate-Severe/Refractory	21
	None	189
Child-Pugh	A	73
	B	36
	C	7
ECOG	0	179
	1	102
	2	29
	≥ 3	10
Extrahepatic invasion	No	74
	Yes	3
Hepatic encephalopathy	Mild/Suppressed	48
	None	120
	Severe/Refactory	3
Macrovascular invasion	No	168
	Yes - major branch	24
	Yes - minor branch	10
Metastasis invasion	No	141
	Yes - distal	7
	Yes - regional	8
Portal hypertension	No	16
	Yes	138
Tumor morphology	Massive, $\geq 50\%$ liver	26
	Multinodular, $< 50\%$ liver	56
	Uninodular, $< 50\%$ liver	132
Tumor number	Single	139
	2-3	47
	> 3cm	26
Tumor size	< 3cm	100
	3-5 cm	63
	> 5 cm	57

Table 4: Text annotation statistics

5.1 Data sparsity for severe conditions

Not all values for each parameter label are well populated in our dataset, as shown in Table 4. Typically the more severe cases are less represented in our data. This was probably due to the nature of our exclusion criteria (only new patients were included), as well as the rapidly declining nature of liver cancer. Five year survival rate is less than 20%, with late-stage patients having less than a year to live (American Cancer Society, 2014). Thus, patients diagnosed at more advanced stages may not be referred to the liver tumor clinic. In our system, we will have to handle these cases of class imbalance.

5.2 Overlapping evidence

Studying our annotations, we observed that related stage parameter types may be referenced by the same text evidence. For example, *"Lesion in seg-*

ment 4A measuring 3.9 x 3.6 cm" implies both that there is a single tumor number and a tumor size between 3-5 cm. Similarly, *"Extrahepatic metastatic disease: None"* suggests both that there is no extrahepatic invasion and no metastasis. Knowledge that some parameters may be grouped into the same evidence may be useful when building the system in terms of joint classification or high-level features. Table 5 gives the groupings of the various stage parameters. Parameters in the same group are more likely to have overlapping evidence, though portal hypertension and macrovascular invasion tend to have little overlap with other evidence types. Ascites and hepatic encephalopathy sometimes reference the same passage, e.g. *"no evidence of liver disease sequelae."* Meanwhile, tumor size, morphology, and number rarely do not reference the same text.

Liver/liver disease	Ascites ChildPugh Hepatic_encephalopathy Portal_hypertension
Overall health	ECOG
Tumor	Tumor_morphology Tumor_number Tumor_size
Liver/liver disease AND tumor	Extrahepatic_invasion Macrovascular_invasion Metastasis

Table 5: Stage parameter groupings

5.3 Explicit vs. non-explicit ECOG evidence

In our annotations, we observed a distinction between text annotation evidence that explicitly mentions an ECOG performance status and those that do not. We define text annotation evidence for a stage parameter as **explicit:** if ECOG (Eastern Cooperative Oncology Group) performance status or any of its abbreviations and acronyms are mentioned in the text evidence.

For example, a text annotation highlight for ECOG, e.g. *"ECOG performance 0,"* is considered explicit. Meanwhile, another ECOG reference, e.g. *"He is cachetic. He is deconditioned and needs a wheelchair to walk greater than 10 feet,"* is considered non-explicit. Other non-explicit mentions may consider patient vocation and habits, e.g. *"He continues to work full time as a security officer"* or *"He lives alone and cares for himself without difficulty."*

To get a sense about the complexity of our task, we divided our ECOG text annotation evidence into explicit and non-explicit evidence by iteratively creating rules and manually inspecting the classification. We found that 170 patients out of the 200 non-excluded patients had some mention of ECOG and as much as 23% of these patients have only non-explicit mentions. Because this division is quite dramatic, we plan to build a separate extraction system for explicit and non-explicit ECOG evidence.

5.4 Missing, ambiguous, and conflicting Child-Pugh evidence

From our 200 patient cohort, only 91 patients had some textual mention of Child-Pugh class. This will necessitate that over half of patients shall require Child-Pugh class calculated according to the logic in Table 1. Accurate Child-Pugh identification will then depend on correct extraction of ascites and hepatic encephalopathy variables. Further complicating the issue, we found cases of ambiguity, e.g. *"He has well-compensated liver disease, with Child-Pugh score of 6 or 7 [...] This puts him at a class A/B"* and cases where separate patient documents gave different Child-Pugh scores. After our final patient level annotations, we can evaluate whether calculated versions of Child-Pugh match with the notes' versions.

5.5 Tumor characteristic reference resolution

We observed for our notes, references to tumor characteristics were often equivocal. Not only were there temporal references to disambiguate, e.g. tumor information from previous readings, but also tumors were identified from radiology artifacts such as "lesions," not all of which were actual tumors. Table 6 shows an example in which 3 lesions are found but only one was suspected to be a HCC tumor. Thus, true tumor size and num-

Focal lesions: Total number: 3: Lesion 1: segment 4A, cm 6.3 x 7.1 X 6.4 cm ... peripherally located lesions are noted in segment 6 measuring 7 mm .. another ... Impression: ... One focal lesion in the segment 4a measuring 6.3 x 7.1 x 6.4 cm. This lesion does not have a typical appearance, but is ... highly suggestive of HCC. Exception: 2 smaller lesions noted in segment 6 peripherally, measuring 11 and 13 mm ... are likely to be arteriovenous shunting.

Table 6: Radiology report excerpt

bers will depend on resolving which lesions are actually tumors, as well as handling reference resolution and temporal factors. Tumor morphology additionally must reason about multiple tumors.

5.6 Discussion

Our inter-rater experiment showed that text annotations are being consistently captured with patient level agreement of 0.91 F1 and partial text span level agreement of 0.73 F1.

A limitation to our process is that most of our phase I text annotations were single annotated. Moreover, we assumed that specific text span passages may be attributed with a label and value assignments however some parameters may require a more patient level holistic view. Furthermore, for our study we focused on patients with available laboratory parameters in structured form. This is not always the case when patients are referred by outside organizations.

Although our annotation phase II has not been fully explored here, we have been able to characterize some of the characteristics in our text annotation evidence, which will inform our extraction task. When our patient level annotations are completed, our multi-level annotation will allow us to run several experiments, including: Given gold patient level stage parameters, how well can a system classify staging? Given gold text level stage parameters, how well can a system predict patient level stage parameters?

6 Machine learning baselines

Once our annotation phase I was completed, we wanted to gauge the complexity of our stage parameter information extraction task. To do so, we created a simple document classification baseline to identify information from phase I text annotations. We chose this baseline because of our sparse annotation approach, i.e. a single document may have several occurrences of the same value but may be annotated only once. Our findings from these experiments will be used to advise us of reasonable performance results and issues to consider for our final system.

6.1 Data

The full corpus of patients was randomly divided into a 20% test, 80% training set. The 160 patient training set included 439 documents (243 clinical, 196 radiology) and 1681 text annotations. The test set will be used in a future comparison of the full staging system against a human abstractor. The training set of patients was divided into 5 folds for training and evaluation.

6.2 Methods

Document level classification was performed for each label-value, e.g. ***ascites-none***. The gold standard document label was automatically inferred by the text annotations from annotation phase I (i.e. If **document0.txt** has been highlighted for ***ascites-none***, then **document0.txt** is marked positive for the gold standard in that classification). The classification was binary, since multiple values for the same label may appear in a single document.

Each label-value document classification only classified document types as prescribed by annotation guidelines. For example, tumor size is restricted to classifying radiology document types, since it is possible they appear in clinical notes but are not annotated due to annotation guidelines. Therefore, ascites label-values classifications occurred over all documents (439 documents), Child-Pugh, ECOG, and hepatic encephalopathy classifications were performed over clinical notes only (243 documents), and the remaining label classifications were on radiology notes only (196 documents).

The features included lower-cased unigram, bigram, and trigram counts after tokenization with punctuations removed. We tested four algorithms with default configurations: C4.5 decision tree, discrete-variable decision tree, and maximum entropy from MALLET (McCallum, 2002), and a linear kernel SVM, scaled by min/maximum values, from LibSVM (Chang and Lin, 2011).

6.3 Results

Results are shown in Table 7. The overall classification performance was 0.63 micro-F1 with the highest and lowest F1 at 0.83 and 0.00, respectively. Best performances per label typically came from the highest frequency class. The best classifiers were the two decision trees, however each classifier was the best in at least one classification.

6.4 Discussion

Analysis of the best-performing baseline models revealed some common limitations. One was the inability to capture long-range logical constructions. One example is for ***ascites - none*** and ***metastasis - no***, which often has passages with long-range negation of related terms. And, as

Label	Freq.	Value	Class.	P	R	F1
Ascites	44	Mild	C45	0.24	0.18	0.21
	20	Moderate-Severe	DT	0.50	0.30	0.38
	146	None	DT	0.77	0.36	0.49
ChildPugh	53	A	DT	0.46	0.49	0.47
	25	B	C45	0.84	0.64	0.73
	7	C	DT	0.50	0.14	0.22
ECOG	105	0	C45	0.71	0.71	0.71
	65	1	DT	0.85	0.54	0.66
	18	2	C45	0.89	0.44	0.59
	8	≥ 3	DT	0.25	0.13	0.17
Extrahepatic invasion	59	No	SVM	0.81	0.85	0.83
	2	Yes	≈	0.00	0.00	0.00
Hepatic encephalopathy	34	Mild	DT	0.70	0.76	0.73
	95	None	DT	0.71	0.73	0.72
	1	Severe	≈	0.00	0.00	0.00
Macro-vascular invasion	127	No	NB	0.71	0.96	0.82
	20	Yes-major_branch	C45	0.50	0.55	0.52
	8	Yes-minor_branch	C45	1.00	0.50	0.67
Metastasis	108	No	DT	0.78	0.70	0.74
	6	Yes-distal	DT	0.50	0.17	0.25
	7	Yes-regional	≈	0.00	0.00	0.00
Portal hypertension	5	No	≈	0.00	0.00	0.00
	84	Yes	C45	0.84	0.80	0.82
Tumor morphology	23	Massive	DT	0.37	0.30	0.33
	40	Multinodular, <50%	ME	0.50	0.15	0.23
	105	Uninodular, <50%	NB	0.62	0.80	0.70
Tumor number	112	Single	NB	0.64	0.84	0.73
	32	2-3	DT	0.24	0.25	0.25
	19	>3	ME	0.67	0.11	0.18
Tumor size	82	< 3	ME	0.64	0.62	0.63
	45	3-5	C45	0.43	0.27	0.33
	46	>5	ME	0.59	0.28	0.38
ALL	1551			0.66	0.60	0.63

Table 7: Best baseline performances for training set. (Freq = frequency of positive cases, Class = classifier, C45 = C4.5 decision tree, DT = binary decision tree, ME = maximum entropy, NB = naive Bayes, SVM = support vector machine)

mentioned in Section 5.5, tumor characteristics require reasoning over several sentences.

Another problem was that these simplistic baseline models and features had difficulty normalizing variations in less frequent equivalent evidence. For example, "abdominal distension" and "abdominal girth" both define ascites but neither term is as frequent as "ascites" so did not become strong features. Similarly, many less frequent Child-Pugh acronyms and abbreviations were missed.

Our baselines also lacked the ability to incorporate outside or domain knowledge to infer information. For example, text evidence for ***ascites - none*** and ***hepatic encephalopathy - none*** can be *"he has no known liver disease,"* which requires knowledge that ascites and hepatic encephalopathy are liver disease symptoms. Non-explicit mentions of ECOG, as discussed in Section 5.3, fall under this category as well. There were also cases in which different values for the same label had very similar language, requiring domain knowledge to differentiate. For example, for macrovascular invasion, while, *"There is thrombus in the right posterior branch of the portal vein [...] possibly [...] tumor thrombus"* is considered ***yes - minor_branch***, *"There is enhancing tumor thrombus in the right portal vein."* is ***yes - major_branch***.

Label-value parameters with higher performances, often harbored strong n-gram features. For example "lactulose," a drug to treat hepatic encephalopathy, was found to be used as an early decision point for both ***mild*** and ***none*** values. For portal hypertension, besides "hypertension", "splenomegaly," spleen enlargement often due to portal hypertension, was a top feature.

Some strategies to overcome current limitations are to use medical ontologies and statistical feature selection to identify terms of interest, which can help normalize for term variations. To handle long-range within-sentence relations, we will apply assertion or negation classifiers and use dependency tree parses to build more complex features. For multi-sentence problems such as the tumors, we will use tools for coreference resolution and time parsing. Furthermore, to reduce noise, we may consider using sub-document level classifications, e.g. at the sentence level.

7 Conclusions and Future Work

In our paper, we described our detailed annotation process, carried out an inter-annotator agreement experiment, and analyzed some of the domain challenges and characteristics of our liver cancer patient data. We were further able to present document classification baselines and analyze their performance. In future work, we will improve information extraction over our current baselines by using targeted feature-rich approaches. We will also extend our system to patient level cancer staging, compare results against a human abstractor, and analyze the affects of using multi-levels of gold input. For example, we may experiment with predicting stages using document-level features vs. extracted text level parameter features.

Although we focus on liver cancer, our workflow may be generalizeable to other cancer or phenotype identification annotation tasks. Futhermore, successful liver cancer parameter identification may be useful for other liver cancer staging schemes or other phenotype cohorts.

References

Alex Vianey Callado França, Jorge Elias Junior, BLG Lima, Ana L C Martinelli, and Flair Jose Carrilho. 2004. *Diagnosis, staging and treatment of hepatocellular carcinoma*, volume 37. Brazilian Journal of Medical and Biological Research.

American Cancer Society 2014. *Cancer facts & figures 2014.* Atlanta, (www.cancer.org/research/cancerfactsstatistics/).

American Cancer Society 2014. *Ovarian Cancer.* (www.cancer.org/cancer/ovariancancer/detailedguide/ovarian-cancer-staging).

Andrew Kachites McCallum. 2002. *MALLET: A Machine Learning for Language Toolkit.* (http://mallet.cs.umass.edu).

Anni Coden, Guergana Savova, Igor Sominsky, Michael Tanenblatt, James Masanz, Karin Schuler, James Cooper, Wei Guan, and Piet C. de Groen. 2009. *Automatically extracting cancer disease characteristics from pathology reports into a disease knowledge representation model*, volume 42. Journal of Biomedical Informatics.

Anthony Nguyen, Darren Moore, Iain McCowan, Mary-Jane Courage. 2007. *Multi-class classification of cancer stages from free-text histology reports using support vector machines*, volume 2007. Annual International Conference of the IEEE Engineering in Medicine and Biology Society.

Anthony Nguyen, Michael J Lawley, David P Hansen, Rayleen V Bowman, Belinda E Clarke, Edwina E Duhig, and Shoni Colquist. 2010. *Symbolic rule-based classification of lung cancer stages from free-text pathology reports*, volume 17. Journal of the American Medical Informatics Association: JAMIA.

Chaitanya Shivade, Preethi Raghavan, Eric Fosler-Lussier, Peter Embi, Nomie Elhadad, Stephen Johnson, Albert Lai. 2014. *A review of approaches to identifying patient phenotype cohorts using electronic health records.*. Journal of the American Medical Informatics Association (JAMIA).

Chih-Chung Chang and Chih-Jen Lin. 2011. *LIBSVM : a library for support vector machines*, volume 2. ACM Transactions on Intelligent Systems and Technology.

David Martinez and Yue Li. 2011. *Information extraction from pathology reports in a hospital setting*, volume 2011. Proceedings of the 20th ACM International Conference on Information and Knowledge Management.

Hui Wang, Weide Zhang, Qiang Zeng, Zuofeng Li, Kaiyan Feng, Lei Liu. 2014. *Extracting important information from chinese operation notes with natural language processing methods types*, volume 48. Journal of Biomedical Informatics.

Iain McCowan, Darren Moore, Anthony N Nguyen, Rayleen V Bowman, Belinda E Clarke, Edwina E Duhig, and Mary-Jane Fry. 2007. *Collection of cancer stage data by classifying free-text medical reports.*, volume 14. Journal of the American Medical Informatics Association: JAMIA.

Jonathan C Yau, Arlene Chan, Tamina Eapen, Keith Oirourke, Libni Eapen. 2002. *Accuracy of the oncology patients information system in a regional cancer centre.*, volume 9. Oncology Reports.

Katherine A McGlynn, W Thomas London. 2011. *The Global Epidemiology of Hepatocellular Carcinoma: Present and Future.*, volume 15. Clinics Liver Disease.

Kwang-Hyub Han, Masatochi Kudo, Sheng-Long Ye, Jong Young Choi, Roonni Tung-Ping Poon, Jinsil Seong, Joong-Won Park, Takafumi Ichida, Jin Wook Chung, Pierce Chow, and Ann-Lii Cheng. 2011. *Asian Consensus Workshop Report: Expert Consensus Guideline for the Management of Intermediate and Advanced Hepatocellular Carcinoma in Asia.*, volume 81. Oncology.

Naveen Ashish, Lisa Dahm, Charles Boicey. 2014. *University of California, Irvine-Pathology Extraction Pipeline: the pathology extraction pipeline for information extraction from pathology reports.*, volume 20. Health Informatics Journal.

Phil Hughes 2000. *Python and Tkinter Programming*, volume 2000. Linux J.

Pontus Stenetorp, Sampo Pyysalo, Goran Topi, Tomoko Ohta, Sophia Ananiadou and Jun'ichi Tsujii. 2012. *brat: a Web-based Tool for NLP-Assisted Text Annotation.* In Proceedings of the Demonstrations Session at EACL 2012.

Ramakanth Kavuluru, Isaac Hands, Durbin Eric B, and Lisa Witt. 2013. *Automatic extraction of icd-o-3 primary sites from cancer pathology reports.* In AMIA Jt Summits Transl Sci Proc.

Tracy Sexton, George Rodrigues, Ed Brecevic, Laura Boyce, Denise Parrack, Michael Lock, David D'Souza. 2002. *Controversies in prostate cancer staging implementation at a tertiary cancer center*, volume 13. The Canadian Journal of Urology.

Vianney Jouhet, G Defossez, Anita Burgun, P le Beux, P Levillain, and Pierre Ingrand. 2012. *Automated Classification of Free-text Pathology Reports for Registration of Incident Cases of Cancer.*, volume 51. Methods of Information in Medicine.

William K Evans, Juanita M Crook, D Read, J Morriss, and DM Logan. 1998. *Capturing tumour stage in a cancer information database.*, volume 2. Cancer prevention & control.

Xiao-Ou Ping, Yi-Ju Tseng, Yufang Chung, Ya-Lin Wu, Ching-Wei Hsu, Pei-Ming Yang, Guan-Tarn

Huang, Feipei Lai, Ja-Der Liang. 2013. *Information extraction for tracking liver cancer patients statuses: from mixture of clinical narrative report types*, volume 19. Telemedicine Journal and E-Health: The Official Journal of the American Telemedicine Association.

Ying Ou and Jon Patrick. 2014. *Automatic Population of Structured Reports from Narrative Pathology Reports types*, volume 163. Proceedings of the Seventh Australasian Workshop on Health Informatics and Knowledge Management. Darlinghurst, Australia.

Yongyut Sirivatanauksorn and Chutwichai Tovikkai. 2011. *Comparison of Staging Systems of Hepatocellular Carcinoma*, volume 2011. HPB Surgery.

Predicting Continued Participation in Online Health Forums

Farig Sadeque
Dept. of Computer and Information Sciences
University of Alabama at Birmingham
Birmingham, AL 35294-1170
farigys@uab.edu

Thamar Solorio
Dept. of Computer Science
University of Houston
Houston, TX 77204-3010
solorio@cs.uh.edu

Ted Pedersen
Dept. of Computer Science
University of Minnesota, Duluth
Duluth, MN 55812-3036
tpederse@d.umn.edu

Prasha Shrestha
Dept. of Computer Science
University of Houston
Houston, TX 77204-3010
pshrestha3@uh.edu

Steven Bethard
Dept. of Computer and Information Sciences
University of Alabama at Birmingham
Birmingham, AL 35294-1170
bethard@uab.edu

Abstract

Online health forums provide advice and emotional solace to their users from a social network of people who have faced similar conditions. Continued participation of users is thus critical to their success. In this paper, we develop machine learning models for predicting whether or not a user will continue to participate in an online health forum. The prediction models are trained and tested over a large dataset collected from the support group based social networking site dailystrength.org. We find that our models can predict continued participation with over 83% accuracy after as little as 1 month observing the user's activities, and that performance increases rapidly up to 1 year of observation. We also show that features such as the time since a user's last activity are consistently predictive regardless of the length of the observation period, while other features, such as the number of times a user replies to others, decrease in predictiveness as the observation period grows.

1 Introduction

Online social networks have established themselves as an integral part of the human interaction in the 21st century. Along with the most popular online social networking services like Facebook, Google+ and the micro-blogging website Twitter, there are other online social networks that are tailored to fit more specific purposes. Among them are the support group based social networks that provide help to individuals with physical or mental afflictions through the sharing of personal experiences and expert advice on a single platform.

Though many aspects of online social networks have attracted the attention of researchers, there is little research to date on computational assessments of engagement among users of online health forums. These forums provide us with information that is unique to this kind of social network: the networks are largely based on the emotional support among the users, so being able to successfully track a user's engagement on these services has the potential for greater impact than the more general social networks. If a support group platform can accurately predict when a user is thinking about leaving, they can take targeted actions to make a more favorable environment for the user, and thus maintain consistent emotional support for their other users. This kind of support is key the health and well-being of the users.

Predicting user engagement is related to concept of *churn prediction* in telecommunication networks (Ngonmang et al., 2012; Mozer et al., 1999; Das-

Proceedings of the Sixth International Workshop on Health Text Mining and Information Analysis (Louhi), pages 12–20,
Lisbon, Portugal, 17 September 2015.

gupta et al., 2008) (predicting when a user will leave one service provider for another), where the motivation is that winning a new customer is more expensive than retaining an existing customer (Hadden et al., 2007). Similarly, retaining an engaged user in a health forum is likely to be easier than engaging a new user in the group. However, users of online support groups are not typically moving between providers, but rather deciding whether or not to continue to use an online support group at all.

In this work, we make the following contributions:

1. We develop models that observe a user for a single month and can predict whether the user will continue participating in the support group in the future with more than 83% accuracy, and can identify users that will leave the group with more than 88% precision and 80% recall.

2. We show that performance on predicting continuing participation rises as the observation period grows beyond one month, rising sharply up to nine months, and then more gradually up to 24 months.

3. We demonstrate a variety of features that are important for this prediction task and show that how often a user replies to others and the time elapsed since their last support group activity are some of the strongest features.

4. We find that the relative importance of the different features changes over time.

We believe this is the first work to look in detail at predicting engagement as it evolves over time in online health forums.

In this work, we focus on the contents of a user's posts and replies and their timeline of activities, rather than the network of friendship links (which is sparse in forum-based social networks as compared to friendship-based social networks like Facebook). We also focus on *active participation*, such as initiating a thread or posting a reply, and not on *passive participation*, such as simply viewing the forum (since such passive information is only available to administrators of the support group service).

2 Data

Our data is collected from DailyStrength[1], one of the largest support group based online social networks with more than 500 support groups based on the physical and mental ailments of its users. Users in these support groups can either *post*, creating a new thread on a new topic, or they can *reply* to a thread that someone else has created. We focused on 20 support groups: Acne, ADHD (Attention Deficit Hyperactivity Disorder), Alcoholism, Asthma, Back Pain, Bipolar Disorder, Bone Cancer, COPD (Chronic Obstructive Pulmonary Disease), Diets and Weight Maintenance, Fibromyalgia, Gastric and Bypass Surgery, Immigration Law, Infertility, Loneliness, Lung Cancer, Migraine, Miscarriage, Pregnancy, Rheumatoid Arthritis and War in Iraq.

We crawled all of the posts (thread initiations) and replies (to existing threads) for these support groups from the earliest available post until the end of September 2013. The posts and replies were downloaded as HTML files, one per thread, where each thread contains an initial post and zero or more of replies. The HTML files were parsed and filtered for scripts and navigation elements to create XML files containing only the users, dates, posts and replies. Each extracted post and reply was part-of-speech tagged using the Stanford part-of-speech tagger (Manning et al., 2014) and was tagged for emotion words by matching it against the Linguistic Inquiry and Word Count (LIWC)[2] lexicon. We also collected the user profile pages of all the users who took part in any form of activity in any of these 20 support groups. Finally, we filtered out the users with the most incomplete profiles, where they were missing both age and gender. These users do not appear in the train, development or test sets, but the replies they post on other users' posts who are not filtered out contribute to the participation prediction task of those users. We also filtered out the user *DS*, the only administrative user in DailyStrength.

Table 1 provides an overview of the resulting dataset. The largest support group among the 20 is Gastric and Bypass Surgery with 21507 posts and 158020 replies and the smallest is the Bone Cancer support group with only 40 posts and 51 replies. The amount of individual activity also varies greatly as there are people who posted or replied only once in their lifetime, and there are

[1] `http://www.dailystrength.org`
[2] `http://www.liwc.net/`

Support groups	20
Posts	110316
Replies	788119
Users	39905

Table 1: Summary of the data collected from DailyStrength

people who have more than 5000 posts or replies.

3 Model

Our goal is to build a model that can observe the past activity of a user on the forum and predict whether the user will continue to participate in the forum in the future. Formally, we would like to construct a model:

$$m_{\Delta t}(u) = \begin{cases} 1 & \text{if } \exists a \in A: \quad a.u = u \wedge \\ & \qquad\qquad\quad a.t > u.t + \Delta t \\ 0 & \text{otherwise} \end{cases}$$

where u is a user, Δt is an amount of time which we call the *observation period*, A is the set of all activities (from any user at any time) such as posting or replying to a post, $a.u$ is the user whose activity it was, $a.t$ is the time of the activity, and $u.t$ is the time at which the user account was created. Intuitively, m should predict 1 iff Δt time has elapsed since the user created their account and there is any new participation (posting or replying) any time in the future after that.

We treat this as a supervised classification problem, and represent a user based on all of his/her activities in the forum during the observation period:

$$rep_{\Delta t}(u) \stackrel{\text{def}}{=} \{a \in A : a.u = u \wedge a.t \leq u.t + \Delta t\}$$

The following sections describe classifier features that are derived from this representation.

3.1 Activity features

These features gather information of a user's activity on DailyStrength. In general, we would expect users who are more active during the observation period to also be more likely to continue to participate in the future.

PostCount The number of threads a user has initiated on the DailyStrength website over the observation period.

ReplyCount The number of replies a user has posted to other users' posts on the DailyStrength website over the observation period.

SelfReplyCount The number of replies a user has posted to their own posts over the observation period.

OtherReplyCount The number of replies a user has received to their posts from other users over the observation period.

3.2 Time features

These features provide a look into the timing of a user's participation on DailyStrength. In general, we would expect users who are participating frequently throughout the observation period to be more likely to participate in the future.

TimeGap1 The number of days between the point at which the user created their DailyStrength account and their first activity (post or reply). This is a measure of how long it took a user to start actively participating in the community.

TimeGap2 The number of days from the time of the last post or reply of a user to the end of the observation period. This is a measure of how long the user has been idle since their last activity.

AvgDays The average number of days between any two sequential activities (posts or replies) by the user during the observation period. This is a measure of how often a user is idle.

3.3 Personal features

These features are gathered from a user's account information page. Since providing age, gender, location and a profile photo are all optional during the DailyStrength account creation process, many users are missing one or more of these pieces of information. In general, we would expect users with more complete profiles to be more likely to continue to participate.

Age The user's age.

Gender The user's gender, either *male*, *female* or *unknown*.

HasLocation A binary feature representing whether or not the user has provided their location.

HasImage A feature representing whether or not the user has provided a profile photo, and if provided, whether it is a stock photo or a user-provided one.

3.4 Content features

These features examine the content of the text in the posts and replies of a user. In general, we would expect users with longer posts to be more likely to continue to participate that users with short posts.

PosUnigrams The total number of words over the observation period that were identified as positive emotions by the LIWC lexicon.

NegUnigrams The total number of words over the observation period that were identified as negative emotions by the LIWC lexicon.

TotalUnigrams The total number of words a user posted over the observation period. This includes all the words (including stop words), not only the emotion words.

Question The total number of questions the user has asked over the observation period in either posts or replies. Questions were identified by looking for sentences ending in question marks.

Url The total number of URLs a user has posted over the observation period.

4 Experiments

For all of the following experiments, we divided the users in our corpus into train, development and test sets with a 60-20-20 ratio, that is, we used 60% of the users to train our prediction model, 20% of the users as the development set and the remaining 20% of the users to test the model. Users were partitioned into each of these sets randomly. These sets do not change with the changes in the observation period, thus giving us the opportunity to compare the results from all observation periods.

We trained classifiers based on WEKA v3.6.11 (Hall et al., 2009), a widely used machine learning toolkit. We normalized all of the aforementioned features to the range [0, 1] to make feature weights more comparable and interpretable. We initially explored several classifiers: naïve Bayes, logistic regression, support vector machines and J48 (a decision-tree based classifier). Logistic regression outperformed the other three in evaluations on the development set, so all results reported here use logistic regression.

We rely on several different performance measures to evaluate our models. First, we report simple classification *accuracy*, the fraction of users for which we correctly predicted whether they would continue participating or leave the forum. We compare this to the *baseline accuracy*, the accuracy of a model that predicts that no users will continue to participate in the forum, and we report *error reduction* of the model accuracy relative to this baseline accuracy. We also report performance measures on the task of identifying users who will not participate in the future: *precision*, the fraction of the users that our model predicted would stop participating who did in fact stop, *recall*, the fraction of the users known to have stopped participating that our model predicted would stop, and *F-measure*, the harmonic mean of precision and recall. Formally:

$$Accuracy = \frac{TP + TN}{TP + TN + FP + FN} \quad (1)$$

$$Precision = \frac{TP}{TP + FP} \quad (2)$$

$$Recall = \frac{TP}{TP + FN} \quad (3)$$

$$F\text{-}measure = 2 \times \frac{Precision \times Recall}{Precision + Recall} \quad (4)$$

where TP is the number of *true positives* (users the model predicted would stop participating and did in fact stop), TN is the number of *true negatives* (users the model predicted would continue participating and did in fact continue), FP is the number of *false positives* (users the model predicted would stop participating but actually continued participating) and FN is the number of *false negatives* (users the model predicted would continue participating but actually stopped participating).

4.1 Can continued participation be predicted?

Our first research question is whether a user's continued participation on the forum can be predicted given the features we developed in Section 3. To test this, we consider an observation period of 1 month and train and test the corresponding classifier. The first row of Table 2 shows the results. Our model achieves 83.06% accuracy, compared to the 50.48% accuracy of the baseline model. For the task of identifying just those users that have

Period	Baseline	Accuracy	Error Reduction	Precision	Recall	F-measure
1	50.48	83.06	65.81	88.3	80.2	84.0
3	63.53	85.69	60.76	92.0	86.4	89.1
6	72.01	87.69	56.02	94.7	89.0	91.7
9	77.75	89.12	51.10	94.9	91.4	93.1
12	82.19	90.71	48.01	96.3	92.7	94.5
15	85.09	92.03	46.54	97.1	93.8	95.4
18	86.96	92.29	40.87	97.3	94.0	95.6
21	87.65	92.34	37.98	97.7	93.8	95.7
24	87.84	92.34	37.01	97.8	93.7	95.7

Table 2: Performance across different observation periods (Period, in months), in terms of baseline accuracy (Baseline, %), model accuracy (Accuracy, %), error reduction of the model over the baseline (ErrRed, %), precision (%), recall (%) and F-measure (%). Precision, recall and F-measure are on the task of identifying users who will not participate in the future.

stopped participating, we achieve 88.3% precision and 80.2% recall. These high performance numbers suggest that while our models are still imperfect, our features are capturing a large proportion of the information necessary to predict continued participation.

4.2 How long must a user be observed?

Our second research question aims to determine the optimal observation period for predicting continued participation. For this experiment, we created 9 observation periods: 1 month, 3 months, 6 months, 9 months, 12 months, 15 months, 18 months, 21 months and 24 months. We then evaluated models trained on these different evaluation periods to see how performance increased or decreased.

Table 2 shows the results. Model accuracy always rises as the observation period grows longer, ranging from 83.06% at 1 month to 92.34% at 24 months. However, the biggest gains are in the shorter periods, with the model increasing accuracy by 7.65% between 1 and 12 months, but only by 1.63% between 12 and 24 months. The performance of the baseline model also increases with the size of the observation period, so that after 24 months 87.84% of all users will not return.

For the task of identifying just those users that have stopped participating, we observe that precision and recall also both rise as the observation period grows, with precision making moderate gains, from 88.32 at 1 month to 97.8 at 24 months, and recall making larger gains, from 80.20 at 1 month to 93.7 at 24 months. As with accuracy, the biggest gains are between 1 and 3 month observation periods.

Overall, these results suggest that observing a user for even 1 month gives reasonable performance, observing for 12 months gives noticeably better performance, and observing for longer than 12 months gives diminishing returns.

4.3 Which features are most important?

Our third research question aims to prioritize our features based on how useful they are to the task of predicting continued participation. To investigate this, we turn to the coefficients (weights) for the independent variables (features) in our logistic regression, which represent the importance of each variable in the classification model. The larger the absolute value of the coefficient, the bigger the impression of that variable on the output. The sign indicates positive or negative effect of that variable on the result, where a negative value means that the feature is associated with continued participation, while a positive value means that the feature is associated with stopping participation.

Table 3 shows the weights of the features obtained from the test data for a 1-month observation period. The most important features (the features with the highest absolute values) are the number of times the user has replied to other users (ReplyCount), the time since the user's last activity (TimeGap2), the time between creating a DailyStrength account and the user's first post (TimeGap1) and the content (Unigram) features. The least important features are mostly the ones aimed at measuring completeness of the profile (Age, Gender, etc.), suggesting that profile completeness is not a good predictor of continued participation. However, the presence of a profile photo

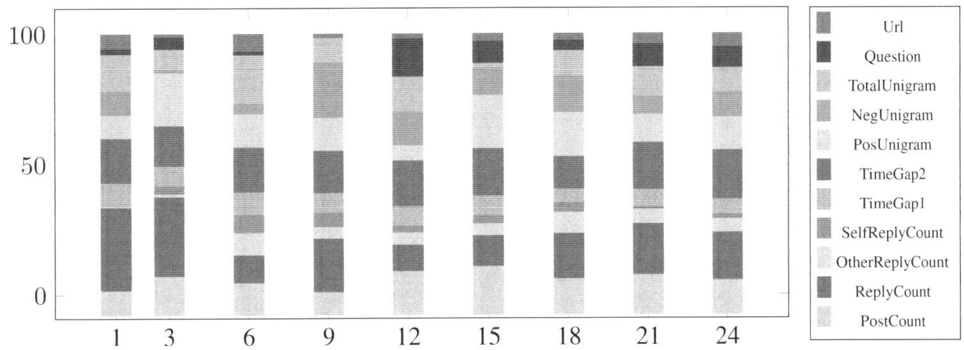

Figure 1: Relative importance of features over the different observation periods. The height of a bar segment represents the absolute value of the weight of the feature, scaled so that the sum of the feature weights is 100%.

Feature	Weight
ReplyCount	-10.652
TimeGap2	5.727
TotalUnigram	4.772
TimeGap1	3.172
NegUnigram	-3.069
PosUnigram	3.002
Url	1.834
HasImage	-0.845
Question	0.827
AvgDays	0.809
PostCount	0.526
Age	-0.346
Gender=unknown	0.202
Gender=female	-0.110
Gender=male	0.093
HasLocation	-0.092
OtherReplyCount	-0.051
SelfReplyCount	0.001

Table 3: Weights of the features for a 1-month observation period

(HasImage) did make a small contribution to the model.

The signs of the weights of the features reveal the direction of predictiveness. The TimeGap1 and TimeGap2 weights are positive, indicating that longer gaps between activities predict someone leaving the forum. PostCount is positive while ReplyCount is negative, suggesting that people who only post will likely leave the forum, while people who reply to others will likely stay. Posting questions and URLs are associated with leaving the forum, along with higher usage of positive unigrams, while higher usage of negative unigram is associated with continued participation.

4.4 Does feature importance change over time?

Our fourth research question asks whether the importance of features is consistent across all observation periods, or whether some features become more or less important than others as the observation period grows.

Figure 1 shows the percentage importance of the eleven most significant features over the different observation periods. Features like TimeGap1 and TimeGap2 are fairly stable in importance over time, with TimeGap1 accounting for 5-9% of the weight and TimeGap2 accounting for 12-18%. ReplyCount is a very strong feature, accounting for as much as 30% in the 1 and 3 month observation periods, but it receives a lower weight for longer observation periods (as little as 10% in the 12 month period). SelfReplyCount and OtherReplyCount, which had almost no weight in the 1 month model, increase in importance for longer observation periods. The other features have less consistent patterns. For example, content features (TotalUnigram, NegUnigram, PosUnigram, Question, Url) account for around 40% of the model weights for most observation periods, but the distribution of weight across these 5 features is erratic over time.

As another measure of feature importance over time, Table 4 shows the increase in accuracy over the baseline majority class model for models trained using only a single feature. Note that the baseline model's accuracy increases for longer observation periods (because more users leave), so

OP	BL	PC	RC	OR	SR	TG1	TG2	AvD	Age	Gen	Loc	Img	Pos	Neg	TUn	Que	Url
1	50.48	4.91	6.55	2.97	2.13	0.00	31.7	4.72	0.36	-0.20	0.00	11.9	5.10	4.32	4.98	4.03	0.10
3	63.53	3.05	3.75	2.43	1.42	0.00	20.7	0.00	0.14	0.00	0.00	0.00	3.06	2.65	2.73	2.52	0.30
6	72.01	1.53	1.89	1.40	0.77	0.00	14.9	0.00	0.00	0.00	0.00	0.00	1.60	1.43	1.50	1.46	0.11
9	77.75	0.81	1.03	0.74	0.24	0.00	10.8	0.00	0.00	0.00	0.00	0.00	0.73	0.72	0.82	0.83	0.07
12	82.19	0.13	0.29	0.11	-0.13	-0.20	7.76	-0.20	-0.20	-0.20	-0.20	-0.20	0.15	0.09	0.13	0.03	-0.18
15	85.09	0.43	0.47	0.37	0.27	0.15	6.16	0.14	0.14	0.14	0.14	0.14	0.33	0.25	0.35	0.36	0.25
18	86.96	0.23	0.20	0.18	0.05	0.00	4.92	0.00	0.00	0.00	0.00	0.00	0.15	0.10	0.09	0.13	0.02
21	87.65	0.08	0.14	0.00	0.00	0.00	4.43	0.00	0.00	0.00	0.00	0.00	0.07	0.07	0.09	0.19	0.02
24	87.84	0.08	0.19	0.01	-0.01	0.00	4.42	0.00	0.00	0.00	0.00	0.00	0.08	0.07	0.09	0.20	0.01

Table 4: Accuracy gain over baseline (BL) over observation periods (OP) when a classifier is trained using only a single feature: PostCount (PC), ReplyCount (RC), OtherReplyCount (OR), SelfReplyCount (SR), TimeGap1 (TG1), TimeGap2 (TG2), AvgDays (AvD), Age, Gender (Gen), HasLocation (Loc), HasImage (Img), PosUnigram (Pos), NegUnigram (Neg), TotalUnigram (Tun), Question (Que), Url

the absolute gains over the baseline always correspondingly decrease. TimeGap2 (TG2) always gives the largest increase in accuracy on its own, as much as 31.7% at a 1 month observation period, and is the only feature that continues (by itself) to give gains over the baseline all the way out to 24 months. ReplyCount (RC) is the next best feature by itself, achieving 6.55% improvement over the baseline at a 1 month observation period, but dropping to less than a 1% improvement by 12 months. The content features PosUnigram (Pos), NegUnigram (Neg), TotalUnigram (TUn) and Question (Que) each achieve a 4-5% improvement over the baseline for a 1 month observation period, but drop below a 1% improvement by 9 months. The personal features generally achieve very little on their own, except for HasImage (Img), which is very useful at 1 month (giving a 11.9% improvement), but giving no improvement for any other observation period.

5 Related Work

Though we are not aware of other models that can observe an online support group user over time and predict their continued participation, there are several works analyzing related problems in other types of social networks. Ngonmang et al. (2012) has worked on a similar problem on a French online blog network called Skyrock[3]. They neither used the contents in the users' posts, nor analyzed the users' behavior. Rather, they used the friendship relationship among the users to predict future participation. There has also been some works on Usenet newsgroups (Joyce and Kraut, 2006; Arguello et al., 2006) where the models take a single post of a user and predict whether or not it will receive a reply. Significant predictors for this task include whether or not the message is cross-posted, the topical coherence of the message with the newsgroup, whether the user posts a question, whether the user is a newcomer to the newsgroup, and the use of third person pronouns. Lampe and Johnston (2005) have shown the effect of feedback on a new user in Slashdot, a news and discussion site. They calculated the user's first comment score and likelihood of getting a second comment by the user based on the feedback the comment received from the other users. They also introduced the time gap between two activities of a user as an indicator of socialization. Mahmud et al. (2014) analyzed word usage to predict social engagement behavior in Twitter. They used psycholinguistic word categories from LIWC, and showed how these categories influence reply and retweet behavior of users. Chen and Pirolli (2012) have analyzed factors that influenced users' engagement in the Occupy Wall Street movement. Danescu-Niculescu-Mizil et al. (2013) has used linguistic change as a predictor of user lifespan in social networks. They conducted their experiments on BeerAdvocate and RateBeer and attempted to find the lifespan of a user early in his or her career.

There are also some studies that include DailyStrength as a data source. Wiley et al. (2014) examined the characteristics of ten different online social networking sites to find impacts of these characteristics on the discussions of pharmaceutical drugs among the users and DailyStrength was one of these websites. Sarker et al. (2015) performed

[3]http://www.skyrock.com

a study on automatic monitoring of Adverse Drug Reactions (ADRs) using user-posted data on social media, and DailyStrength was a data source for their study.

6 Discussion

Our findings have several implications for social interaction in online health forums. This is the first study that attempts to predict continued participation of users in such support groups. Though the model is not perfect, it produces results with high accuracy, precision and recall. The high precision and recall has greater significance in this experiment, as they represent our model's correctness in identifying the people who leave the group after a certain observation period. Identifying these people early in their lifecycle will help social health platforms identify users that are not being fully served, allowing the platforms to analyze the reason for the departure and create a more favorable environment for everyone.

This is also the first study that examines the effect of different lengths of observation period to determine the minimum amount of time required to accurately predict future participation. With a 12-month observation period, we can predict continued engagement with high accuracy, precision and recall, though even at a 1-month observation period, performance is good.

Our work has shown which features contribute the most to predict a user's continued participation. As we can see from the results, personal features covering demographics and profile completeness play little to no part in predicting user's engagement, whereas the other three categories have varied significance over time. The predictiveness of time based features, especially the time from account creation until a user's first activity and time since a user's last activity, are consistently predictive over all lengths of observation. The predictiveness of replies to other users' posts is very large for 1 and 3 month observation periods, but is a little less informative for larger observation periods. The predictiveness of content features (word count, negative/positive words, etc.) is generally good, though which of these features is most important varies somewhat over time.

Although the model we built produces high performance results, there are several opportunities to improve it further. As our results show that the usage of positive and negative unigrams has some influence over the prediction task, we plan to expand these features by using additional psycholinguistic word categories to find other relations between emotional status and continued participation. We also plan to expand beyond word unigrams, which fail to account for phenomena such as negation (where *good* becomes *not good*), incorporating longer linguistic dependencies into these features. And we plan to use more linguistic features to capture explicit speech acts in the posts of the user that may indicate their intent to leave the forum. In addition, we plan to analyze the replies received by other users to find out whether there is a pattern that encourages or discourages a user's participation, such as a long duration between posting and receiving a reply, or the use of harsh or aggressive forms of language. Finally, we plan to explore machine learning formulations that will allow us to dynamically extend our observation period for a user to just the point at which we can confidently predict whether or not they will continue to participate.

7 Conclusion

In this paper we presented a study to determine what makes a user continue his or her participation in a support group based online health forum and how long we have to observe a user's activity to predict this accurately. We built a model that predicts continued participation of a user and we showed that this model is accurate with an observation period as little as one month. Increasing the observation period increases performance, though most of the gains are achieved by the end of 12 months. The model reveals that features like the time since a user's last activity and the number of times a user has replied to others are consistently strong predictors of continued participation. Our model forms a foundation for future research in modeling the evolution over time of user engagement in online health forums.

Acknowledgments

We would like to thank Binod Gyawali who did the initial data gathering and data coding.

References

Jaime Arguello, Brian S. Butler, Elisabeth Joyce, Robert Kraut, Kimberly S. Ling, Carolyn Rosé, and Xiaoqing Wang. 2006. Talk to me: Foundations for successful individual-group interactions in online communities. In *Proceedings of the SIGCHI Conference on Human Factors in Computing Systems*, CHI '06, pages 959–968, New York, NY, USA. ACM.

Jilin Chen and Peter Pirolli. 2012. Why you are more engaged: Factors influencing twitter engagement in occupy wall street. In *Sixth International AAAI Conference on Weblogs and Social Media*.

Cristian Danescu-Niculescu-Mizil, Robert West, Dan Jurafsky, Jure Leskovec, and Christopher Potts. 2013. No country for old members: User lifecycle and linguistic change in online communities. In *Proceedings of the 22Nd International Conference on World Wide Web*, WWW '13, pages 307–318, Republic and Canton of Geneva, Switzerland. International World Wide Web Conferences Steering Committee.

Koustuv Dasgupta, Rahul Singh, Balaji Viswanathan, Dipanjan Chakraborty, Sougata Mukherjea, Amit A Nanavati, and Anupam Joshi. 2008. Social ties and their relevance to churn in mobile telecom networks. In *Proceedings of the 11th international conference on Extending database technology: Advances in database technology*, pages 668–677. ACM.

John Hadden, Ashutosh Tiwari, Rajkumar Roy, and Dymitr Ruta. 2007. Computer assisted customer churn management: State-of-the-art and future trends. *Computers & Operations Research*, 34(10):2902–2917.

Mark Hall, Eibe Frank, Geoffrey Holmes, Bernhard Pfahringer, Peter Reutemann, and Ian H. Witten. 2009. The weka data mining software: An update. *SIGKDD Explor. Newsl.*, 11(1):10–18, November.

Elisabeth Joyce and Robert E. Kraut. 2006. Predicting continued participation in newsgroups. *Journal of Computer-Mediated Communication*, 11(3):723–747.

Cliff Lampe and Erik Johnston. 2005. Follow the (slash) dot: Effects of feedback on new members in an online community. In *Proceedings of the 2005 International ACM SIGGROUP Conference on Supporting Group Work*, GROUP '05, pages 11–20, New York, NY, USA. ACM.

Jalal Mahmud, Jilin Chen, and Jeffrey Nichols. 2014. Why are you more engaged? predicting social engagement from word use. *CoRR*, abs/1402.6690.

Christopher Manning, Mihai Surdeanu, John Bauer, Jenny Finkel, Steven Bethard, and David McClosky. 2014. The stanford corenlp natural language processing toolkit. In *Proceedings of 52nd Annual Meeting of the Association for Computational Linguistics: System Demonstrations*, pages 55–60, Baltimore, Maryland, June. Association for Computational Linguistics.

Michael Mozer, Richard H Wolniewicz, David B Grimes, Eric Johnson, and Howard Kaushansky. 1999. Churn reduction in the wireless industry. In *NIPS*, pages 935–941.

Blaise Ngonmang, Emmanuel Viennet, and Maurice Tchuente. 2012. Churn prediction in a real online social network using local community analysis. In *Proceedings of the 2012 International Conference on Advances in Social Networks Analysis and Mining (ASONAM 2012)*, ASONAM '12, pages 282–288, Washington, DC, USA. IEEE Computer Society.

A. Sarker, A. Nikfarjam, K. O'Connor, R. Ginn, G. Gonzalez, T. Upadhaya, S. Jayaraman, and K. Smith. 2015. Utilizing social media data for pharmacovigilance: A review. *J Biomed Inform*, Feb.

M. T. Wiley, C. Jin, V. Hristidis, and K. M. Esterling. 2014. Pharmaceutical drugs chatter on Online Social Networks. *J Biomed Inform*, 49:245–254, Jun.

Redundancy in French Electronic Health Records: A preliminary study

Eva D'hondt
LIMSI-CNRS UPR 3251
Rue John von Neuman
91403 Orsay, France
dhondt@limsi.fr

Xavier Tannier
LIMSI-CNRS UPR 3251
Univ. Paris-Sud
Rue John von Neuman
91403 Orsay, France
xtannier@limsi.fr

Aurélie Névéol
LIMSI-CNRS UPR 3251
Rue John von Neuman
91403 Orsay, France
neveol@limsi.fr

Abstract

The use of Electronic Health Records (EHRs) is becoming more prevalent in healthcare institutions world-wide. These digital records contain a wealth of information on patients' health in the form of Natural Language text. The electronic format of the clinical notes has evident advantages in terms of storage and shareability, but also makes it easy to duplicate information from one document to another through copy-pasting. Previous studies have shown that (copy-paste-induced) redundancy can reach high levels in American EHRs, and that these high levels of redundancy have a negative effect on the performance of Natural Language Processing (NLP) tools that are used to process EHRs automatically. In this paper, we present a preliminary study on the level of redundancy in French EHRs. We study the evolution of redundancy over time, and its occurrence in respect to different document types and sections in a small corpus comprising of three patient records (361 documents). We find that average redundancy levels in our subset are lower than those observed in U.S. corpora (respectively 33% vs. up to 78%), which may indicate different cultural practices between these two countries. Moreover, we find no evidence of the incremental increase (over time) of redundant text in clinical notes which has been found in American EHRs. These results suggest that redundancy mitigating strategies may not be needed when processing French EHRs.

1 Introduction

Electronic Health Records (EHRs) are becoming prevalent in most healthcare institutions and have been recognized to contain crucial information about patients' health in the form of Natural Language text. As a result, specialized Natural Language Processing (NLP) methods and tools are being developed to unlock this wealth of medical information from EHRs and use it in medical applications such as clinical decision support (Demner-Fushman et al., 2009). The electronic format of the clinical notes in the patient records makes it easy to duplicate information from one document to another through copy-pasting methods. Previous studies (Wrenn et al., 2010; Zhang et al., 2011; Cohen et al., 2013) have shown that the amount of redundancy introduced by copy-pasting could reach up to 78% in clinical notes from American hospitals. This situation makes it difficult to exploit the content of EHRs both for humans and NLP tools (Cohen et al., 2013).

One motivation for introducing content redundancy in clinical documents is a need for completeness: While a patient's history grows over time with each new hospital visit, the key information remains the same. By copy-pasting all available information into the most recent document, this becomes a stand-alone document which offers a complete and up-to-date overview of the patient's history and status. Other reasons for the observed redundancy are more pragmatic: logistical issues such as the time to load previous documents from the EHR or restricted access rights to documents created in a different hospital department can lead health professionals to ensure that all relevant information is present in the clinical note they are currently writing.

Proceedings of the Sixth International Workshop on Health Text Mining and Information Analysis (Louhi), pages 21–30,
Lisbon, Portugal, 17 September 2015.

These practices can make health professionals more efficient but they may also represent potential risks to patient care by creating confusion between what is relevant to the patient's current versus his or her past medical condition (Siegler and Adelman, 2009; Weis and Levy, 2014). Interestingly, in the process of copying information from an older document into a new one, small changes can be introduced into the narrative, such as typo corrections, acronym expansion, and information updates. For these reasons it is important to study the nature and extent of redundancy in clinical notes in more detail (Zhang et al., 2014). An important goal is to identify which portions of a clinical document are entirely new and which portions are redundant from previous documents in the records, and whether the redundant portions are identical to previous documents, or modified.

To our knowledge, the issue of redundancy in Electronic Records has been predominantly studied for English, and more specifically in documents produced in healthcare institutions in the United States. In this paper, we present a preliminary study that addresses redundancy in French clinical narratives from a group of healthcare institutions in France. We analyse a corpus of 361 documents from 3 patient records and examine to what extent and under which conditions redundancy is present.

2 Objective and research questions

The goal of this study is to characterize redundancy in French EHRs with a view to gauge its impact on NLP processing. This work is motivated by the findings of similar studies on US EHRs, which have shown that a certain level of redundancy (30% and more) affects word distribution in the documents and therefore has a (negative) impact on language models. Also, when annotating data, redundancy could lead to duplicate work, which we do want to avoid.

It is important to note that our study focuses exclusively on *surface redundancy*, which we define as text sections of a document that are copied verbatim (or with marginal edits) to another document. While surface redundancy, i.e. literal copying, entails redundant information, the reverse is not necessarily true: In a patient's records, the same information may be repeated in various documents but using different wording. Although this type of paraphrasing may be considered as conveying redundant information, it does not have a negative impact on language models, because the utterances are part of the natural language diversity of expression.

In this paper we aim to answer the following research questions:

- Does surface redundancy grow over time in a patient's records?
- Which parts of the documents contain the most redundancy?
- Are certain types of documents in EHRs more likely to contain redundant information than others?
- Is there more redundancy within patient's records versus between patient's records?

We expect the results of this study to provide some insight on the suitable natural language processing methods to apply to French EHRs. In particular, in light of the research conducted on US clinical notes, we need to understand the nature of redundancy in French data in order to decide whether redundancy mitigation strategies are needed.

3 Background

Redundancy detection is closely related to the research topic of plagiarism detection, but there are some key differences between the two fields. Since redundancy is introduced by (indiscriminate) copying of text without much human intervention, redundancy detection is mainly focused on literal string matching, rather than employing semantic similarity measures (other than detecting spelled-out variants of acronyms) or paraphrase detection. Furthermore, redundancy detection is usually performed within a closed reference collection (as opposed to plagiarism detection systems that use the entire internet as a reference base).

It is important to note that near-duplicate blocks of texts that are copied from a source text can occur in different positions in the new document. As a result, similarity measures that treat the whole document as one string i.e., global alignment, are not optimal for redundancy detection (see also discussion in Zhang et al. (2011)).

Table 1 presents an overview of the tools and methods whose suitability for redundancy detection in our corpus we reviewed in the course of this study. The Baldr[1] and Sherlock[2] (Mozgovoy et al., 2005) software packages have been developed for plagiarism detection exclusively. Baldr is a source-code plagiarism-detecting software that uses 'information distance' (Vitányi et al., 2009) to measure similarity between two documents. The intuition underlying this distance is that two objects (in this case text documents) are similar if the transformation function to transform one document into the other is simple to describe. If, however, all such functions are complex, the objects are deemed dissimilar. Baldr uses real-world compression software to calculate the transformation metrics (Chen et al., 2004). The Sherlock software package uses the more common method of fingerprinting, i.e. hashing substrings of the text into unique digital signatures. Redundancy is then calculated as the proportion of common signatures between an incoming document and documents in the comparison set. Unlike the method described below for Cohen et al. (2013), the Sherlock program operates on word level, i.e. uses words as its basic units, instead of characters.

The other methods in Table 1 were developed specifically for the task of redundancy assessment in EHRs. They were all applied on corpora from healthcare institutions in the United States. Cohen et al. (2013) developed a character-based fingerprinting method similar to the BLAST sequence similarity method (Altschul et al., 1990) which is popular in bioinformatics. When applied to a subset of a large corpus of 22,500 patient notes, they observed an average level of redundancy of around 30% within patient records, but a much lower amount of redundancy (on average 2.9%) between patient records. They also found that redundancy in a large corpus has a significant negative effect on the performance of language modelling applications.

Wrenn et al. (2010) developed a token-based Levenshtein edit distance measure to perform sequence alignment between two documents. The reported redundancy score is the proportion of aligned tokens over the total number of tokens in the base document. In their study they looked at the occurrence of redundancy over time in a corpus of 100 EHRs (admissions) and found levels of redundancy between 54% and 78% depending on the clinical note (i.e. document) type. Furthermore they noted that the level of redundancy consistently increased over time in the corpus.

Zhang et al. (2011) used vector-based semantic similarity measures to measure redundancy in outpatient notes. They analysed a corpus of notes from 178 patients and found that these notes contain a large amount of redundancy. Like Wrenn et al. (2010) they also studied time progression, and observed that note redundancy increased over time.

4 Material and Methods

4.1 Corpus

For this study, we used a set of French clinical notes where personally identifying information (PII) had been marked and replaced by surrogates (Grouin and Névéol, 2014). The documents were also marked with four types of content sections: letterhead, patient header, content and footer (Deléger et al., 2014). One of our goals is to assess whether there is more redundancy in notes belonging to one patient, compared to redundancy in notes across patient records. To this end we selected three complete patient records for our corpus. These records contain a total of 361 documents. Each record comprised of at least 100 documents and tracks the treatment of a patient over the course of several years. To allow for a fair comparison of redundancy within and between patient records, we selected three patient records with similar profiles, i.e. patients that were admitted for renal transplantation and follow-up care.

4.2 Measuring redundancy

For comparability with previous work, we measured corpus redundancy using the fingerprinting method (Cohen et al., 2013). We also developed our own fingerprinting method, which is an extension of the Cohen method: Like the Cohen measure our implementation calculates a similarity score based on the proportion of n-character fingerprints which a target document has in common with a base document or collection, over the total number of fingerprints in the target document. In other words, it shows what proportion of the text (expressed in fingerprints) is redundant, i.e. has also appeared in the base document. Unlike Cohen's method our implementation allows for the extraction of overlapping fingerprints, which im-

[1]http://wassner.blogspot.fr/2014/05/baldr-loutil-anti-fraude-anti-plagiat.html

[2]http://www2.warwick.ac.uk/fac/sci/dcs/research/ias/software/sherlock/

name	method	score range	time-ordering	comparison
Cohen et al.	non-overlapping fingerprints	0-1	no	document-pairs
adaptedCohen	overlapping fingerprints	0-1	yes	document-pairs, corpus
Wren et al.	Levenstein distance	0-1	yes	corpus
Zhang et al.	semantic similarity	0-1	yes	corpus
Baldr	compression range	0-1	no	document-pairs
Sherlock	overlapping fingerprints	0-100	no	document-pairs

Table 1: Overview of redundancy measuring tools reviewed; the tools specifically used in this work appear in *italic* font.

proves both coverage of the original text and allows for a more precise calculation of the number of fingerprints that are in common. It is also robust against differences in lower/uppercase, insertion of spaces and newlines. For the analyses reported in this paper, we converted the whole document to a single string and extracted overlapping fingerprints of 30 characters with 10-character intervals. (In the original Cohen implementation non-overlapping 30-character fingerprints are extracted line per line.) Since we are interested in the temporal aspects of patient records, our script takes timestamps of documents into account which allows for chronological sorting and comparison between individual documents as well as that of a document to the concatenation of all older documents in the corpus.

4.3 Vizualization of redundancy

In section 6 we describe a prototype system for the visualization of patient records and how it can be used for annotation purposes. The code underlying the prototype is a Python wrapper script that takes temporally-ordered document-pair redundancy scores from the adapted Cohen script and uses this information to dynamically generate a graph (using the GraphViz software package) which depicts the flow of information in the patient's records over time.

5 Results

5.1 Incremental redundancy

The three subfigures in Figure 1 show the progression of redundancy over time in each of the three patient records in our corpus, measured with the Cohen script, and our own adapted Cohen script, respectively. Each data point shows the proportion of redundant text in a given document (ticks on the x-axis), compared to the concatenation of text from all older documents in the corpus. The documents on the x-axis are ordered chronologically. Since the original Cohen script does not allow for sequential comparison implementation, we performed manual data selection of the older documents to ensure that the two implementations were tested on the same corpus subsets.

While similarity measures should not be directly compared, they both show similar evolutions in the patient's records. We see that there is no clear incremental growth of redundancy over time such as has been reported for American EHRs, in any of the patient records. It should be noted that patient records 3 shows an increase in redundancy scores for the 20 most recent (i.e. right-hand) documents. Closer analysis shows that this is likely due to the type of documents, namely discharge notes (*Compte Rendu de Séjour*). The level of redundancy in different document types is discussed below.

We find that although both measures show very similar progressions, the adapted Cohen script allows for a more precise measuring than the original Cohen script (as evidenced by the higher average redundancy scores in all three subfigures). We will therefore be using this implementation for the other analyses reported in the rest of the paper.

5.2 Comparison between sections

While the previous analysis showed that there is no incremental growth of redundancy in the corpus, redundancy is still present: On average[3], 33% of the text in a document in the corpus is redundant. To estimate the impact on text mining and determine whether it will be beneficial or harmful, it is important to characterize which parts of the document are more likely to contain redundant information. To this end we created a second version of the corpus in which the header and footer information for all documents had been removed,

[3] Calculated with adapted Cohen script

Category	Records 1	Records 2	Records 3[4]
CR d'Acte	10.3 (14.1)	16.3 (21.2)	15.9 (23.8)
CR de Séjour	25.7 (25.9)	N/A[5]	38.0 (30.5)
TA de Séjour	7.3 (9.9)	9 (15.9)	9.8 (12.3)

Table 2: Average redundancy for different document types. All numbers are percentages. Between brackets is the standard deviation. 'CR' stands for 'Compte Rendu' (*report*), 'TA' stands for 'Text Associé' (*associated text*).

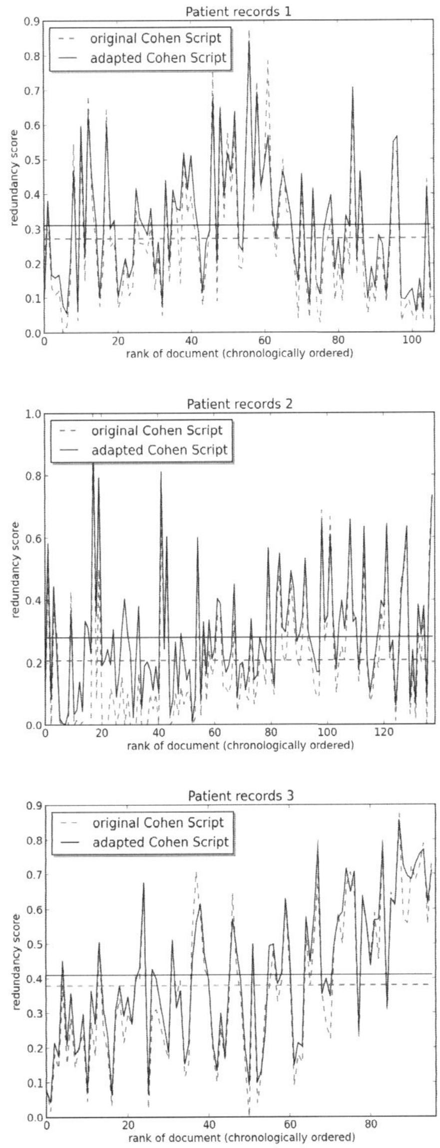

Figure 1: Redundancy over time of original text in three patient records, calculated by the Cohen and adaptedCohen script. The flat lines show the average over the whole patient records.

leaving only 'topical content'.

Figure 2 shows the same progression of redundancy over time as Figure 1, but calculated over the version of the corpus without header or footer information. We see that the overall trends of the respective graphs remain similar for the different patient records but that the average redundancy level has decreased drastically. In patient records 1, 2 and 3, the average redundancy level decreases from 31% to 17.8%, 28% to 15.7% and 41% to 23.3%, respectively.

It is clear that most of the redundant text appears in the header and footer sections of the document. This text is not very informative by nature: Headers and footers contain contact information such as names, addresses, de-identified patient names, which will only add noise for text mining purposes that want to exploit the Natural Language in the EHRs. In the following analyses, we therefore only use the NoHeaders versions of the patient records, that is, only the free text that makes up the body of the notes in the patient records.

5.3 Comparison between document categories

Patient records contain a wealth of information in a variety of document types, such as test results and surgery notes (*Compte Rendu d'Acte*), discharge summaries (*Compte Rendu de Séjour*), and correspondence between doctors from various hospital departments (*Texte associé de Séjour*). As each document type describes a different aspect of the patient's stay in a hospital, they are likely to differ in writing style but also in their purpose in the hospital. Following Wrenn et al. (2010) we studied the differences in redundancies between different document types.

Table 2 shows the differences in average redundancy levels of documents from the three main document categories in the three patient records.

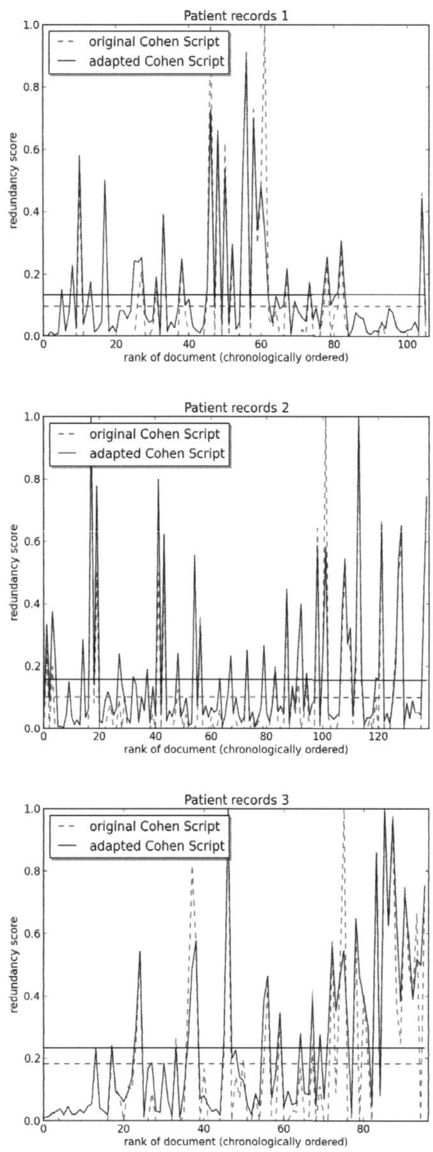

Figure 2: Redundancy over time of text in three patient records with header and footer information removed, calculated by the Cohen and adaptedCohen script. The flat lines show the average over the whole patient records.

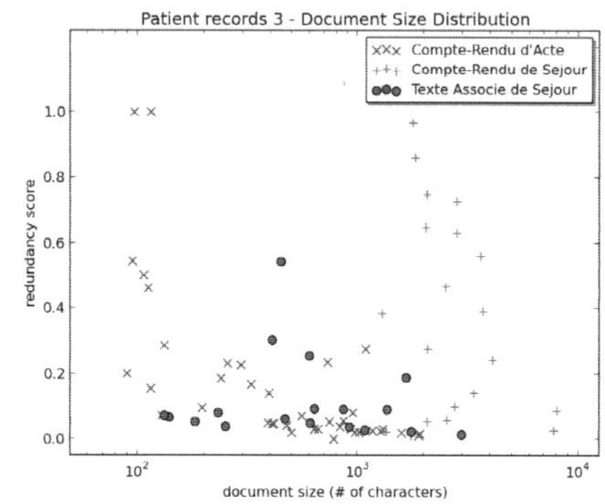

Figure 3: Document size (in number of characters) versus redundancy levels of documents from the main three categories of the patient records of patient 3. X-axis is set to logarithmic scale.

We can see that the average level of redundancy differs substantially between the different document types. Discharge summaries (*Compte Rendu de Séjour*) contain by far the most redundancy, while the associated correspondence between doctors (*Texte Associé de Séjour*) have a fairly low amount of redundancy. This can be explained by their structure and use: Discharge summaries generally have a fixed structure and aim to give a full overview of the patient's stay in the hospital, as well as a short overview of the patient's history. The associated correspondence, however, is optional, and is typically in the form of a letter that contains free text and no fixed structure.

To ensure that the measured redundancy levels are not an artefact of document length, we performed an additional analysis of document size versus redundancy ratio. Figure 3 shows the size distribution (in number of characters) of the documents from the three largest categories in the patient records of patient 3. Patient records 1 and 2 show similar distributions.

We find no direct correlation between document length and redundancy level, but rather a U-shaped distribution. The longest documents documents, i.e. discharge summaries (*Compte Rendu de Séjour*) have the highest average redundancy.

[4] As the metadata provided for these patient records contained errors, the documents have been manually reclassified into the different document types.

[5] The patient records did not contain enough documents of this type to calculate a reliable average

T / S	Records 1	Records 2	Records 3
Records 1	*17.8 (13.5)*	14.4 (18.8)	12.4 (15.1)
Records 2	12.8 (17.9)	*15.7 (20.4)*	15.0 (18.0)
Records 3	9.5 (13.7)	12.1 (18.4)	*23.3 (24.9)*

Table 3: Redundancy scores between different patient records. All numbers are percentages. Between brackets is the standard deviation. S stands for *source* corpus. T stands for *target* corpus. The number in italics are the average redundancy levels that correspond to the black flat lines in Figure 2.

5.4 Inter-/Intra-patient record comparison

We saw that redundancy within patient record is fairly low (compared to the scores reported for the American EHRs), but given the fixed structure of certain document types (*Compte Rendu de Séjour* and *Compte Rendu d'Acte*) phrases or formulations may be shared between different records. Identifying these would be helpful for categorizing similar interventions and tests (as described over different documents) in different patient records.

Table 3 shows that the average redundancy between patient records is slightly lower than those within the EHRs. We can conclude that even though the patient illnesses and treatments (and the associated forms to record these) are fairly similar, the text in the patient records differs substantially between patients.

6 Discussion

Cross-culture differences in generating and managing medical information in text have been studied previously for breast cancer forums in Germany vs. United Kingdom (Weissenberger et al., 2004) and EHRs structure and narrative style for China vs. United States (Wu et al., 2013), and Sweden vs. Finland (Allvin et al., 2011). In this study we offer a preliminary comparison of the occurrence of redundancy in French EHRs, compared to numbers reported on redundancy in EHRs from the United States.

Although our corpus of the three patient records is too small to give conclusive results, it does offer some interesting insights: We find that the level of surface redundancy present in (and between) French medical records is fairly low. Moreover, we do not see clear indications of an incremental increase of redundancy over time, as has been reported for American EHRs. Most of the redundancy that is present in the French records in our study comes from document headers and footers. This text does not offer information on the course of the patient illness and should be discarded so as not to harm the performance of Text Mining or NLP applications that use the EHRs as training material. The absence of redundancy in the body of texts is beneficial for text mining purposes, and we can conclude that the mitigating strategies that have been developed for American EHRs are not probably not needed when processing French EHRs.

Subjective review of some of the French clinical notes confirms our findings and suggests that the copy-paste practices observed in American hospitals, which are meant to give health professionals access to comprehensive information about a patient within a single document, are not used in France. We find that rather than copy-pasting content from previous documents, references to previous documents are inserted in new documents (such as *cf. CR précédent*, see previous report, *examen biologique: voir feuilles ci-joint*, lab results: see attached).

However, as Table 2 indicates, this does differ between category types: In discharge summaries (*Compte-Rendus de Séjour*), which are generally the longest documents in the corpus, the measured redundancy levels are higher than in other types of documents, which indicates that entire text portions are copied from older documents. Discharge summaries are meant to be stand-alone documents integrating information about an entire patient stay which is otherwise described minutely in several other documents. We notice that these copied portions of text are often not strictly copy-pasted as they integrate small differences corresponding to re-writing of the text for clarity, addition of details previously unavailable and correction of erroneous information. So rather than copy-pasting content indiscriminately, French health professionals seem to do it strategically. In this way, redundancy should not be seen as a source of noise in a corpus but rather as an indication of information flow between documents.

These observations suggest that documents containing highly redundant sections are key documents in the patient records. While many of these documents are identified as *Compte-Rendu de Séjour* in the metadata, this is not always the case. Therefore, it would be important to automatically identify these documents in a given EHR, so as to provide a new doctor with the most complete overview of a patient's history. Such information could later also serve for the purposes of automatic summarization.

As an exploration of our hypothesis and to gather more insight into the structure of patient records, we developed the prototype of a visualization tool that would allow us to track how information is transferred in a patient record over time. Figure 4 shows a screen capture of (part of) the graph generated for one of the patient records used in this study. Each block in the figure corresponds to one document in the patient's records. The documents are ordered chronologically along the Y-axis from earliest (top) to most recent (bottom). Documents that were created at the same moment, i.e. during the same hospital stay, are thus positioned next to one another. The shape of the blocks refers to the document type (*Compte-Rendu de Séjour* are square, *Compte-Rendu d'Acte* are circles, ...), and their size is relative to the document size. The number in the block refers to the document identifier in the patient's records. The interlinking lines refer to the proportion of redundant information in the more recent document that comes from the older document. A user-defined cut-off parameter allows for interactive exploration.

In this study we have focused exclusively on surface redundancy, i.e. the (almost) literal repetition of a piece of text from an older document. While string-based similarity measures are useful to detect blatant copy-pasting that throws off word distributions in language models, they prove too crude when we want to detect the flow of information, i.e. strategically used copy-pasting. More specifically, the current method cannot deal with highly similar text that is used to described two different events. For example, blood test results tend to be communicated using the same standard form. If two different blood tests yield the same results, this will result in two highly similar documents, the most recent of which would be judged as highly redundant by our current method. A more precise method is needed that incorporates semantic components such as identification of temporal expressions, or even event detection, into the string similarity method.

7 Future Work

The work presented in this paper is a preliminary study on a small-scale corpus and was meant to gain insight into copy-paste-induced redundancy in French EHRs. We find that rather than focusing on mitigating methods (as needed for American EHRs), we should look toward developing high-precision measures that capitalize on the existing redundancy in French EHRs. A first step for future work will be to replicate this study on a larger and more diverse corpus of patient records with different disease profiles, so as to confirm our findings and see to what extent text is shared between patient records from different hospital departments.

As a follow-up of this study we also plan to address two new main lines of research. First, we intend to develop a more precise surface redundancy measure which takes temporal expressions and terminological variation into account and which is more robust to small changes within large highly similar context. We will use the WiCoPaCo corpus (Max and Wisniewski, 2010) to train models that can automatically identify reformulations, and distinguish those from (error) corrections and updates.

Second, we will study redundancy on the level of the patient's records as a whole, not just on the document level. We intend to develop a measure that uses information on redundancy levels, the number of documents copied, the (temporal) distance of information that has been copied, ... to identify key documents within a patient's record. To this end we will need a reference set of correctly identified key documents in a set of patient records. This will be carried out by a group of health professionals, who will manually classify the documents in a sizable set of EHRs in terms of their importance (and information content). Since such annotations are very expensive, these professionals will be provided with an improved version of the visualization tool described in section 6 to select potential documents and speed up the annotation process.

8 Conclusion

In this paper, we present a preliminary study on the presence of copy-paste-induced redundancy in French EHRs. We find that the high levels of redundancy and incremental increase of redundant text over time which have been observed in American EHRs, does not feature in our subset. As a re-

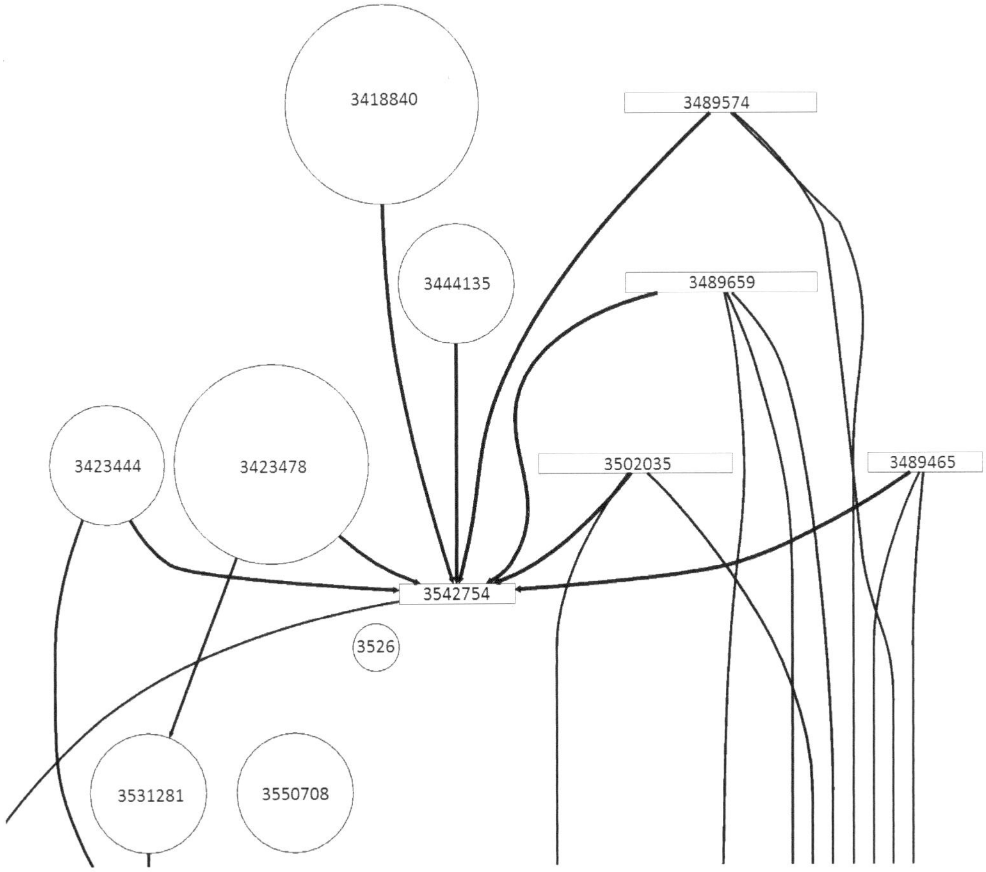

Figure 4: Screen shot of zoom-in from the visualization tool. Documents (represented by blocks) are ordered chronologically (Y-axis) with the oldest at the top, and the most recent at the bottom. The number within a block refer to the document identification number within the respective patient's records.

sult, there is no expected impact from redundancy on language models or other natural language processing methods applied to French EHRs. Rather, the limited redundancy that is present in the corpus may be strategically exploited to yield important information from the records.

Acknowledgments

This work was supported by the French National Agency for Research under grants CABeRneT[6] ANR-13-JS02-0009-01 and Accordys[7] ANR-12-CORD-0007.

The authors thank the Biomedical Informatics Department at the Rouen University Hospital for providing access to the LERUDI corpus for this work.

[6]CABeRneT: Compréhension Automatique de Textes Biomédicaux pour la Recherche Translationnelle

[7]Agrégation de Contenus et de COnnaissances pour Raisonner à partir de cas dans la DYSmorphologie foetale

References

H Allvin, E Carlsson, H Dalianis, R Danielsson-Ojala, V Daudaravičius, M Hassel, D Kokkinakis, H Lundgrén-Laine, GH Nilsson, O Nytrø, S Salanterä, M Skeppstedt, H Suominen, and S Velupillai. 2011. Characteristics of finnish and swedish intensive care nursing narratives: a comparative analysis to support the development of clinical language technologies. *J Biomed Semantics*, Suppl 3:S1.

Stephen F Altschul, Warren Gish, Webb Miller, Eugene W Myers, and David J Lipman. 1990. Basic local alignment search tool. *Journal of molecular biology*, 215(3):403–410.

Xin Chen, Brent Francia, Ming Li, Brian Mckinnon, and Amit Seker. 2004. Shared information and program plagiarism detection. *Information Theory, IEEE Transactions on*, 50(7):1545–1551.

R Cohen, M Elhadad, and N Elhadad. 2013. Redundancy in electronic health record corpora: analysis, impact on text mining performance and mitigation strategies. *BMC Bioinformatics*, 14:10.

L Deléger, C Grouin, and A Névéol. 2014. Automatic content extraction for designing a french clinical corpus. In *Proc AMIA Annu Symp*.

Dina Demner-Fushman, Wendy W Chapman, and Clement J McDonald. 2009. What can natural language processing do for clinical decision support? *J Biomed Inform*, 42:760–772.

Cyril Grouin and Aurélie Névéol. 2014. De-identification of clinical notes in french: towards a protocol for reference corpus developpement. In *J Biomed Inform*, Aug.

Aurélien Max and Guillaume Wisniewski. 2010. Mining naturally-occurring corrections and paraphrases from wikipedia's revision history. In *LREC*.

Maxim Mozgovoy, Kimmo Fredriksson, Daniel White, Mike Joy, and Erkki Sutinen. 2005. Fast plagiarism detection system. In *String Processing and Information Retrieval*, pages 267–270. Springer.

EL Siegler and R Adelman. 2009. Copy and paste: a remediable hazard of electronic health records. *Am J Med*, 122:495–496.

Paul MB Vitányi, Frank J Balbach, Rudi L Cilibrasi, and Ming Li. 2009. Normalized information distance. In *Information theory and statistical learning*, pages 45–82. Springer.

JM Weis and PC Levy. 2014. Copy, paste, and cloned notes in electronic health records: prevalence, benefits, risks, and best practice recommendations. *Chest*, 145(3):632–8, Mar 1.

C Weissenberger, S Jonassen, J Beranek-Chiu, M Neumann, D Müller, S Bartelt, S Schulz, JS Mönting, K Henne, G Gitsch, and G Witucki. 2004. Breast cancer: patient information needs reflected in english and german web sites. *Br J Cancer*, 91(8):1482–7, Oct 18.

JO Wrenn, DM Stein, S Bakken, and PD Stetson. 2010. Quantifying clinical narrative redundancy in an electronic health record. *Journal of the American Medical Informatics Association*, 17(1):49–53.

Y Wu, J Lei, WQ Wei, B Tang, JC Denny, ST Rosenbloom, RA Miller, DA Giuse, K Zheng, and H Xu. 2013. Analyzing differences between chinese and english clinical text: a cross-institution comparison of discharge summaries in two languages. In *Stud Health Technol Inform*, volume 192, pages 662–6.

R Zhang, S Pakhomov, BT McInnes, and GB Melton. 2011. Evaluating measures of redundancy in clinical texts. In *Proc. AMIA Annual Symposium*, page 1612–1620.

R Zhang, S Pakhomov, and GB Melton. 2014. Longitudinal analysis of new information types in clinical notes. In *Proc. AMIA Summits on Translational Science*, page 232–237.

Is it possible to recover personal health information from an automatically de-identified corpus of French EHRs?

Cyril Grouin
LIMSI-CNRS, UPR 3251
Rue John von Neuman
91400 Orsay, France
grouin@limsi.fr

Nicolas Griffon
LIMICS, INSERM U 1142
CISMeF-TIBS-LITIS, EA 4108
CHU de Rouen, 76031 Rouen
nicolas.griffon@chu-rouen.fr

Aurélie Névéol
LIMSI-CNRS, UPR 3251
Rue John von Neuman
91400 Orsay, France
neveol@limsi.fr

Abstract

De-identification aims at preserving patient confidentiality while enabling the use of clinical documents for furthering medical research. Herein, we aim to evaluate whether patient re-identification is possible on a corpus of de-identified clinical documents in French. Personal Health Identifiers are automatically marked by a de-identification system applied to the corpus, followed by reintroduction of plausible surrogates. The resulting documents are shown to individuals with varying knowledge of the documents and de-identification method. The individuals are asked to re-identify the patients. The amount of information recovered increases with familiarity with the documents and/or de-identification method. Surrogate re-introduction with localization from the same (vs. different) geographical area as the original documents is found more effective. The amount of information recovered was not sufficient to re-identify any of the patients, except when privileged access to the hospital health information system and several documents about the same patient were available.

1 Introduction

Research using clinical data requires the informed consent of patients involved. Privacy rules and regulation in France require that, in the absence of informed consent, clinical records used in research be anonymized or de-identified.

Anonymization consists in ensuring that health data used in the research can not be linked to individual patients. Alternatively, de-identification consists in removing or hiding personal health identifiers found in health documents (Meystre et al., 2010). In this study, we focus on the result of an automatic de-identification process. Both anonymization and de-identification aim at preserving patient confidentiality while enabling the use of clinical documents for furthering medical research. State-of-the-art automatic de-identification methods are often evaluated for their ability to redact a set of personal health identifiers (PHI) from clinical documents (Meystre et al., 2010). PHIs are defined according to the American Health Insurance Portability and Accountability Act (HIPAA) of 1996*.

In this study, we are investigating whether it is possible for individuals to recover patients' personal information based on the content of automatically de-identified documents. We caracterize the re-identification attempts using the skills, tools or information at the attacker's disposal. The targets of the re-identification attempts can be both surrogates wrongly used in replacement of original PHI and data not processed during the de-identification step (data missed during the de-identification process as well as data not being in the scope of this process).

Assessing whether patients can be re-identified after documents have been automatically de-identified is a difficult task, since the combination of seemingly innocuous pieces of information could endanger patient privacy (Benitez and Malin, 2010; Barbaro and Zeller Jr, 2006). The combination of a de-identification system that automatically tags PHIs in clinical text with the replacement of PHIs by plausible surrogates has been used to create realistic modified clinical records (Sweeney, 1996; Neamatullah et al., 2008) that are clinically and linguistically valid. This

*U.S. Department of Health Human Services, 1996 http://www.hhs.gov/ocr/privacy/hipaa/administrative/privacyrule/adminsimpregtext.pdf

Proceedings of the Sixth International Workshop on Health Text Mining and Information Analysis (Louhi), pages 31–39, Lisbon, Portugal, 17 September 2015.

method is also referred to as "hiding in plain sight" obfuscation and was shown to contribute to increase the effective recall of automatic de-identification systems from about .94 to .99 (Carrell et al., 2013).

While the impact of de-identification on clinical information contained in the records has been studied (Deleger et al., 2013; Meystre et al., 2014b), fewer efforts have addressed the effective impact on patient privacy. Encouragingly, it was recently shown that doctors were not able to identify patients they had recently treated when relying on de-identified records (Meystre et al., 2014a). There is a need for other studies that evaluate whether re-identification is possible based on de-identified records.

In this study, we address re-identification attempts from the perspective of making a small de-identified clinical corpus available to the research community in circumstances such as a shared task or NLP challenge. Although a dataset released in the context of a shared task or challenge would require participants to sign a user agreement specifically binding recipients to *not* engage in re-identification attempts, this study considers an attack scenario where a negligent (or malignant) user would overlook this requirement.

In this context, we anticipate that the corpus would be accessed by individuals with a variety of backgrounds including researchers, developpers and clinicians. Furthermore, depending on the type of NLP task addressed by the challenge, there may be a need to include several documents pertaining to the same patient (e.g., to evaluate systems that create a cross-document patient timeline) or not (e.g., to evaluate systems that perform named-entity or concept recognition).

Accordingly, we consider re-identification attempts by individuals with varied knowledge of clinical records and de-identification methods (medical doctors and computer scientists) on automatically de-identified records in French. In addition, we also assess the success of re-identification attempts on different types of datasets (documents pertaining to the same patient, vs. random patients) and surrogate re-introduction methods (using localization information similar to that of original documents, vs. different). The corpus used in our study has been automatically de-identified by a system, without any human intervention to check the outputs produced by the automatic process.

2 Background

The release of datasets containing personal information about the individuals who contributed to the creation of the data raises the concern of privacy protection. When such datasets are prepared for research purposes, the risks of privacy breach must be assessed and weighed against the potential benefits the research conducted using the data. In prior instances of data release, inadequate assessment of the possibility of privacy breach has led to public embarrassment and legal action (Barbaro and Zeller Jr, 2006). In light of this experience, extreme caution is needed prior to releasing sensitive data. The case of medical data such as those contained in Electronic Health Records requires specific attention, since the first rule of medical ethics as outlined in the Hippocratic Oath is to "first, do no harm". This makes it unethical to release medical data that could cause harm to a patient, e.g., through privacy breach.

The Multiparameter Intelligent Monitoring in Intensive Care II (MIMIC-II) database (Saeed et al., 2002; Saeed et al., 2011; Lee et al., 2011) is an example of a success story in the clinical domain. In addition to applying a high-performing automatic de-identification method, the creators of MIMIC have drawn a data use agreement that requires the users to be informed about the sensitive nature of the data, and to contribute to privacy protection, should they identify any potential breach. To our knowledge, this is the only clinical database of this scale available for clinical and Natural Language Processing (NLP) research in English or in any other language. Smaller de-identified clinical datasets have also been released in conditions similar to MIMIC in the context of international NLP challenges, such as i2b2 with a variety of goals, including the evaluation of de-identification methods (Uzuner et al., 2007).

We believe that studies assessing the possibility of privacy breach on realistically de-identified data can lead to a better understanding of the risk benefit balance for dataset release. In addition, such studies can contribute to building confidence in de-identification systems and methods that are otherwise evaluated quantitatively.

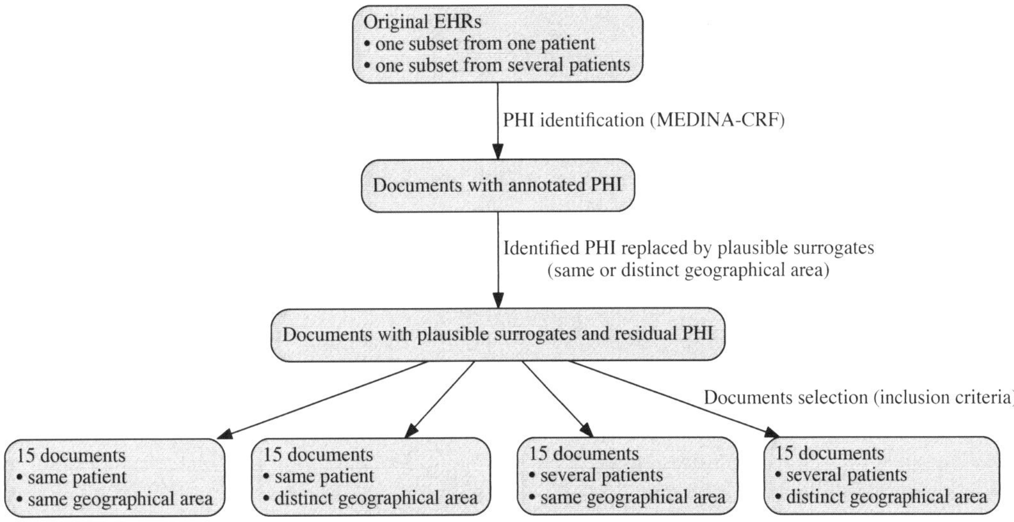

Figure 1: Production of corpora used in this study

3 Material and methods

3.1 Corpus preparation

The corpus used in this study was approved by the French administrative authority on data privacy[†] for research on Information Retrieval (IR) in large Electronic Health Records.

Twelve types of PHIs pertaining to patients, patient relatives and health professionals were targeted in our study: first names, last names, initials, addresses, cities, countries, zip codes, telephone and fax numbers, email addresses, hospital names, identifiers (such as social security numbers or medical device serial number) and dates (including patient date of birth).

In this study, we selected documents among the three most frequent types in the corpus: discharge summaries, correspondance and procedure or consult report.

We assess the chances of re-identification on a worst-case scenario, using a high-performing automatic de-identification method (Medina-CRF, see details below) on a corpus comprising 60 documents that will likely cause the system to fail identifying some PHI.

Rule-based criteria for finding files that we anticipate to be "hard to de-identify" for the automatic tool were compiled based on an error analysis.

[†]CNIL - Commission nationale de l'informatique et des libertés `http:www.cnil.fr`

They include:

Name criteria The tool often fails to identify complex names or part of complex names that include a hyphen or space (e.g., Dorothy Jane, Watterman-Smith).

Contact information criteria The tool often fails to identify contact information that appears in the content section of document (i.e., outside of header/footer sections), even when introduced by trigger words (e.g., "domicilié" *residing at*, "personne de confiance" *support person*)

Date criteria The tool often erroneously marks dates that are not linked to the patient record, e.g., dates of legal procedures quoted in the patient record. Marking these dates for replacement can compromise the confidentiality of the other dates in the file or record, because marked dates are shifted by a random number of days at the step of surrogate re-introduction.

The Medina-CRF de-identification tool for French clinical documents was designed by one of the authors (Developer 2). It is a statistical tool that was trained on a corpus of 100 gold-standard documents (Grouin and Névéol, 2014). The automatically tagged PHIs are replaced by plausible surrogates in order to create a de-identified corpus where PHIs may or may not pertain to the original documents.

The idea behind surrogate introduction is to apply the "hiding in plain sight" principle, with the hypothesis that original PHIs will be less conspicuous in the corpus among surrogate PHIs. The original version of the surrogate replacement module was developed by one of the authors (Developer 2) and then extended by another author (Developer 1).

We assess the possibility of re-identification in different situations, relevant to clinical NLP research using de-identified records. Depending on the aim of a study, it may be necessary to use a corpus comprising documents pertaining to medical record of the same patient (e.g., patient timeline analysis) or documents pertaining to different patients (e.g., concept identification).

Our hypothesis is that re-identification might be more difficult for a corpus of documents pertaining to random patients (vs. same patients), as a corpus of documents from the same patient provides more information about a unique patient and also offers the possibility to cross-reference information between documents.

We also assess the possibility of re-identification with different settings of the surrogate re-introduction tool. Our surrogate re-introduction method relies on lists of surrogates for each type of PHI that can be marked by the de-identification tool. A setting of the tool allows the user to select a geographical area (at the level of French departments, equivalent to U.S. states) for the re-introduction of surrogates for cities, zip codes and hospitals.

We experimented with two settings of the tool, one where the geographical area of surrogates was the same as that of original PHIs, one where the geographical area was different.

The corpus was divided into four sections to study the variations in medical purpose and surrogate setting (see Figure 1):

1. 15 documents pertaining to the same patient with surrogate reintroduction from the same geographical area

2. 15 documents pertaining to random patients with surrogate reintroduction from the same geographical area

3. 15 documents pertaining to the same patient with surrogate reintroduction from a different geographical area

4. 15 documents pertaining to random patients with surrogate reintroduction from a different geographical area

For files collected from the same patient, we selected a random record from all records with at least one file meeting the "hard to de-identify" criteria. Then, we selected one random file meeting the "hard to de-identify" criteria for this record, and then selected other files randomly from the record.

For files collected from random patients we first selected three files meeting each type of criteria, and then selected other files randomly.

The file selection method was automatic, so that the authors who designed the method and participated in the annotation (Developers 1 and 2) knew of the selection criteria, but did not have prior knowledge of why a particular file in the corpus had been selected.

3.2 Gold standard set development

Two sets of gold standard annotations were created for the study corpus of 60 documents.

One gold standard set comprises annotations of all PHIs on the original corpus (see gold standard 1 in Figure 2). It was obtained by revising the original corpus with automatically marked PHIs. This gold standard is used to determine the performance of the de-identification tool on the study corpus.

Another gold standard set comprises annotations of the original PHIs that were not replaced by surrogates (see gold standard 2 in Figure 2). This was obtained by comparing the study corpus (after surrogates were re-introduced) to the original corpus (with automatically marked PHIs). One annotator prepared this gold standard corpus after they had produced their annotations on the original corpus, in an experimental setting similar to that of other annotators. This gold standard set is used to determine the number of "unmarked" PHIs in the study corpus, and to compute the performance of annotators to identify original PHIs.

3.3 Re-identification experiment

The corpus was shown to individuals with varying knowledge of the documents and corpus processing method: one clinician practicing in the hospital that supplied the corpus for this study, two informatics researchers who designed and developed Medina (the de-identification tool and surrogate re-introduction tool), and three other researchers without specific knowledge of the corpus or de-identification methods. Each individual was asked to mark PHIs that they believed to be original, i.e., that may reveal information about the patients. The annotations were made using the BRAT rapid annotation tool (Stenetorp et al., 2012).

The annotators that were not familiar with either the corpus or the de-identification method (Researchers 1-3) were told briefly that clinical documents were processed automatically to replace twelve types of PHIs. They were given the specific list of PHIs, which was encoded in BRAT as categories available for creating annotations. They were told that the automatic system was not perfect, and that some of the PHIs present in the documents might be original PHIs, that they had to try and identify.

The annotators were told that the four sections of the corpus corresponded to a document selection from either the same or random patients. Researchers 1-3 were told that the geographical setting of surrogate re-introduction varied between corpus sections. However, they were not told that one setting was the original geographical location, while the other was not.

After the annotators had worked on the documents, they were asked to provide any specific information on any patient that they believed to have re-identified in the course of the study. They could use any tool at their disposal to attempt re-identifying the patients. In practice, the tools used included: a generic search engine, an online reverse look-up directory service and a hospital health information system.

For each individual, we computed the performance of identifying original PHIs as well as inter-annotator agreement (IAA) with other individuals in terms of F-measure. Performance of PHI identification and IAA were assessed both overall for the entire corpus as well as for each of the four sub-sections.

PHI type	Total	Unmarked
Last name	541	18 (3.3%)
First name	487	17 (3.5%)
Initials	39	35 (89.7%)
Address	60	21 (51.7%)
City	153	39 (25.5%)
Zip Code	67	12 (17.9%)
Phone	282	0 (0.0%)
Email	42	0 (0.0%)
Identifier	20	16 (80.0%)
Date	233	17 (7.3%)
Hospital	166	24 (14.5%)

Table 1: Distribution of total and unmarked PHIs in the final corpus

Category	Precision	Recall	F-measure
Last name	0.97	0.95	0.96
First name	0.98	0.96	0.97
Initials	0.67	0.05	0.09
Identifier	1.00	0.25	0.40
Hospital	0.74	0.53	0.62
Address	0.98	0.82	0.89
Zip code	1.00	0.79	0.88
City/Country	0.99	0.95	0.97
Date	0.94	0.97	0.96
E-mail	1.00	1.00	1.00
Telephone	0.99	1.00	0.99
Overall	0.96	0.90	0.93

Table 2: Performance of Medina-CRF on the study corpus

4 Results

Table 1 shows the distribution of PHIs in the corpus. About 10.0% were original PHIs, while 90.0% were re-introduced surrogates.

Table 2 shows the detailed performance of the automatic de-identification tool, for exact matches. The overall performance on the corpus was 0.93 F-measure, with 0.96 precision and 0.90 recall, which can be considered state-of-the-art.

Figure 2 shows an excerpt of a sample corpus document. This document was selected as "hard to identify" per our contact information criteria as it contains the trigger word "personne de confiance" *support person*, along with a contact phone number for the patient's spouse. While this particular PHI was correctly identified and substituted by the automatic system, additional information about the patient's family was not. Documents are shown to annotators without any markings (processed text). On the gold standard 2 section of the figure, surrogate PHIs are shown in italic font, and original PHIs (that were not substituted by the automatic processing) are underlined. In this example, the original PHIs were the residence location of the patient's children - the passage reports "Marital Status: married. 3 Children (2 in Marseille, 1 in Corse)." For this particular document, two annotators (Developer 1 and 2) correctly identified that the two original PHIs were, indeed, original. One annotator (Researcher 1) identified that

original text	**gold standard 1**
Mary Smith	Mary Smith
née le 05/08/1928	née le 05/08/1928
Mariée, 3 enfants (2 à Marseille et 1 en Corse)	Mariée, 3 enfants (2 à Marseille et 1 en Corse)
Profession: sans profession	Profession: sans profession
...	...
Personne de confiance: époux Tél: 06 41 69 31 72	Personne de confiance: époux Tél: 06 41 69 31 72
...	...
Pathologie pancréatique en 1993	Pathologie pancréatique en 1993
...	...
Dr. Daniel Lucas, Médecin attaché.	Dr. Daniel Lucas, Médecin attaché.
processed text, shown to annotators	**gold standard 2**
Jane Doe	*Jane Doe*
née le 04/07/1927	née le *04/07/1927*
Mariée, 3 enfants (2 à Marseille et 1 en Corse)	Mariée, 3 enfants (2 à Marseille et 1 en Corse)
Profession: sans profession	Profession: sans profession
...	...
Personne de confiance: époux Tél: 06 02 41 57 15	Personne de confiance: époux Tél: *06 02 41 57 15*
...	...
Pathologie pancréatique en 1992	Pathologie pancréatique en *1992*
...	...
Dr. Gregory House, Médecin attaché.	Dr. *Gregory House*, Médecin attaché.

Figure 2: Sample corpus document. Original PHIs (annotated in the gold standard corpora) are underlined. For illustration purposes on this figure, surrogate PHIs are shown in *italic font*.

the country PHI "Corse" was original. Relying on their knowledge of the "hard to identify" criteria, Developer 1 also marked the phone number "06 02 41 57 15" as original PHI, when it was in fact a surrogate.

On average, the annotators each spent 2 hours working on the corpus to produce the annotations.

Table 3 presents the performance of PHI identification by annotator, ordered by prior knowledge of data and method; we can classify them into three groups, represented by double bars: advanced knowledge of both documents and method, advanced knowledge of either documents or method, little knowledge of either documents or method. The table presents results for each of the four sections of the corpus (lines 2 to 5) as well as overall (line 6).

Table 4 presents the inter-annotator agreement for PHI identification.

Patient re-identification using the generic search engine and online reverse look-up directory service was unsuccessful. However, two patients could be re-identified using the hospital health information system.

5 Discussion

Performance of original PHI identification Table 3 shows that overall, PHI recognition is low. It suggests that the ability to identify original PHIs is associated with prior knowledge of the documents and/or corpus de-identification method. The highest PHI recognition is 0.50, which is not very high performance. Researchers 1-3 had no prior knowledge of either the method or the documents. After the experiment, Researcher 1 correctly identified the hospital that supplied the documents. No individual was able to supply more specific information about any of the patients based on the corpus alone.

Table 4 shows that the higher inter-annotator agreement was observed between the annotators with the highest performance for PHI recognition, Developer 1 and Clinician. Nonetheless, agreement was only 0.33, which is considered very low (Artstein and Poesio, 2008). This indicates that the "hiding in plain sight" strategy is working well, and that the original PHIs are not obvious to the annotators.

Corpus	1	2	3	4	Overall	
Dev1	34	35	31	42	142	n
	0.71	0.57	0.61	0.67	0.62	P
	0.33	0.54	0.40	0.50	0.50	R
	0.45	0.56	0.49	0.57	0.57	F
Clin	13	11	19	25	68	n
	0.62	0.64	0.47	0.76	0.61	P
	0.11	0.18	0.19	0.34	0.20	R
	0.19	0.29	0.27	0.47	0.30	F
Dev2	285	59	28	41	413	n
	0.16	0.19	0.71	0.51	0.23	P
	0.64	0.30	0.43	0.38	0.46	R
	0.26	0.23	0.53	0.43	0.30	F
Res1	30	8	6	15	59	n
	0.47	0.50	0.33	0.80	0.54	P
	0.19	0.11	0.04	0.21	0.15	R
	0.27	0.18	0.08	0.34	0.23	F
Res2	0	66	0	43	109	n
	0.00	0.02	0.00	0.07	0.04	P
	0.00	0.03	0.00	0.05	0.02	R
	0.00	0.02	0.00	0.06	0.03	F
Res3	26	24	26	10	86	n
	0.00	0.00	0.00	0.00	0.00	P
	0.00	0.00	0.00	0.00	0.00	R
	0.00	0.00	0.00	0.00	0.00	F

Table 3: Performance of PHI identification in terms of number of PHIs annotated (n), Precision (P), Recall (R) and F-measure (F). Clin=Clinician, Dev=Developer, Res=Researcher. The corpus subsets are listed as per the description in section 3.1: 1=same patient, same location; 2=random patients, same location; 3=same patient, different location; 4=random patients, different location

	Dev1	**Clin**	**Dev2**	**Res1**	**Res2**
Clin	0.32	–			
Dev2	0.21	0.10	–		
Res1	0.21	0.11	0.18	–	
Res2	0.00	0.00	0.00	0.01	–
Res3	0.01	0.01	0.03	0.01	0.00

Table 4: Inter-Annotator Agreement in terms of F-measure (Clin=Clinician, Dev=Developer, Res=Researcher)

Methods for re-identification attempts The tools available to the annotators to attempt re-identifying the patients mainly consisted of information publicly available over the internet.

One annotator (Researcher 1) systematically checked hospital names, person names and locations using a generic search engine, and was able to identify the hospital that the patients were treated in. Two annotators (Developer 1 and Clinician) used an online reverse look-up directory service for all phone numbers and addresses that they believed might be original PHIs; however, the queries did not return any results.

One annotator (Clinician) had access to the hospital Health Information System (HIS). They reported that the information they were able to recover about any single patient in the random sets was likely not sufficient to submit a valid query into the system[‡]. However, for patients with several documents available, recouping an approximate stay date with diagnostic codes such as ICD or CPT either found directly in a document (these codes are not considered PHIs) or inferred using medical coding knowledge, they were able to pull the patient record from the HIS and therefore identify the patients. For the two patients in our study corpus (corpus subsets 1 and 3) the re-identification required several attempts at querying the HIS and took 20 minutes for one patient, 30 minutes for the other.

Other annotators did not report using any re-identification strategies that relied on data sources outside the documents themselves.

The most powerful tool used is definitely the hospital HIS, by an individual with access credentials. The system search services are set-up in a way that requires the users to provide sufficient information about the patients before a record can be retrieved. In our study, it did not help with patient re-identification when there was only one document available about the patient.

However, when there were several documents available about the same patient, the patient could be identified in 30 minutes or less. In this case, patient identification required (i) access to the HIS; (ii) knowledge of how records are coded and retrieved in the HIS; and (iii) medical knowledge to identify or infer diagnostic codes from the patient documents.

[‡] In practice, re-identification was attempted for a few documents that the Clinician thought had re-identification potential. Attempts to re-identify a patient were stopped after 30 minutes when they were not successful. After a few failures, the clinician focused his efforts on the patients for which more than one document were available.

Perfomance variation according to medical purpose and surrogate setting We can see from table 3 that overall, the performance of PHI identification seems to be higher when the geographical area of re-introduced surrogate is different from that of the original PHIs (row 4 vs. row 2, row 5 vs. row 3). This indicates that the "hiding in plain sight effect" is stronger when location surrogates come from the same geographical area as the original PHIs.

The initial argument against using same-area location surrogates was that, as surrogates are re-introduced randomly, a surrogate could happen to be the same as the original PHI, therefore voiding the replacement operation. We did identify a few occurrences of this phenomenon in our data set, when the surrogate and original PHI were in fact different strings (e.g., "Bois-Guillaume" vs. "BOISGUILLAUME").

The PHI identification results from table 3 do not clearly indicate that PHI identification was made easier in the corpus subsets of documents from the same patients vs. random patients (line 2 vs. line 3 and line 4 vs. line 5). However, when the hospital HIS was available, patients with multiple documents available could be identified while patients with only document available could not.

Implications for the design of a clinical corpus to be used in an NLP challenge or shared-task The results of our study suggest that re-identification attempts from researchers without privileged access to the hospital health information system (which is expected to be the case of most individuals accessing a corpus through a challenge) will not be successful.

It is also important to point out that the identification of the patient identities in this study were only possible because the de-identification was performed automatically and some original PHIs (dates) could be found in the documents.

In the context of data release for a challenge or shared-task, the de-identification process should include multiple rounds of manual review of PHI to ensure that no original PHIs were left.

In summary, this study suggests that patient privacy can be reasonably preserved in a corpus comprising documents pertaining to random patients, with same-area geographical surrogate re-introduction and manually reviewed de-identification.

Limitations The main limitation in this study is the size of the corpus, which comprises 60 documents. This size was chosen to keep the annotation time manageable. It is comparable to the size of the corpus (85 documents) used previously by Meystre et al. (2014a). The study of variations leads us to partition the corpus into four subsets of 15 documents, which can only provide indicative results. The study will need to be reproduced on a larger scale.

Also, one important category of individual likely to identify patients from the content of de-identified files includes patients themselves, or patients' relatives and acquaintances. For instance, an individual who personally knows the patient that our sample file pertains to (see Figure 2) might read this document and realize that the information (stay at home mother of 3 children who experienced a pancreas disorder in the past) matches the circumstances of their acquaintance. However, we have not been able to devise an adequate experimental setting to evaluate this chance. Arguably, the chance might be similar to that of patient re-identification by a doctor who had personally attended to the patient within the past three months. It was found that doctors were not able to re-identify their own patients from de-identified documents (Meystre et al., 2014a).

6 Conclusion

In spite of shortcomings of the de-identification system identified by the developpers in a thorough error-analysis, patient privacy was not compromised by individuals without privileged access to the relevant hospital health information system.

When access to the hospital health information system is available, patients can be re-identified by recouping information found in more than one document, and medical knowledge of medical coding. However, patient privacy is preserved when only one document per patient is available.

Furthermore, less information can be recovered when location surrogates for the same geographical area as the original files are used.

Acknowledgement

This work was supported by the French National Agency for Research under grant CABeRneT[§] ANR-13-JS02-0009-01. The authors thank the

[§]CABeRneT: Compréhension Automatique de Textes Biomédicaux pour la Recherche Translationnelle

Biomedical Informatics Department at the Rouen University Hospital for providing access to the LERUDI corpus for this work, and the annotators who kindly participated in this study.

References

Ron Artstein and Massimo Poesio. 2008. Inter-coder agreement for computational linguistics. *Computational Linguistics*, 34(4):555–96.

Michael Barbaro and Tom Zeller Jr. 2006. A face is exposed for aol searcher no. 4417749. *The New York Times*, August 9.

Kathleen Benitez and Bradley Malin. 2010. Evaluating re-identification risks with respect to the HIPAA privacy rule. *J Am Med Inform Assoc*, 17(2):169–77.

David Carrell, Bradley Malin, John Aberdeen, Samuel Bayer, Cheryl Clark, Ben Wellner, and Lynette Hirschman. 2013. Hiding in plain sight: use of realistic surrogates to reduce exposure of protected health information in clinical text. *J Am Med Inform Assoc*, 20(2):342–8, Mar-Apr.

Louise Deleger, Katalin Molnar, Guergana Savova, Fei Xia, Todd Lingren, Qi Li, Keith Marsolo, Anil Jegga, Megan Kaiser, Laura Stoutenborough, and Imre Solti. 2013. Large-scale evaluation of automated clinical note de-identification and its impact on information extraction. *J Am Med Inform Assoc*, 20(1):84–94.

Cyril Grouin and Aurélie Névéol. 2014. De-identification of clinical notes in french: towards a protocol for reference corpus developpement. *J Biomed Inform*, 46(3):506–515, Aug.

Joon Lee, Daniel J Scott, Mauricio Villarroel, Gari D Clifford, Mohammed Saeed, and Roger G Mark. 2011. Open-access MIMIC-II database for intensive care research. In *Proc IEEE Eng Med Biol Soc*, pages 8315–8.

Stephane M Meystre, F Jeffrey Friedlin, Brett R South, Shuying Shen, and Matthew H Samore. 2010. Automatic de-identification of textual documents in the electronic health record: a review of recent research. *BMC Med Res Methodol*, 10(70).

Stephane Meystre, Shuying Shen, Deborah Hofmann, and Adi Gundlapalli. 2014a. Can physicians recognize their own patients in de-identified notes? In *Stud Health Technol Inform*, volume 205, pages 778–82.

Stephane M Meystre, Oscar Ferrández, F Jeffrey Friedlin, Brett R South, Shuying Shen, and Matthew H Samore. 2014b. Text de-identification for privacy protection: a study of its impact on clinical text information content. *J Biomed Inform*, 50:142–50, Aug.

Ishna Neamatullah, Margaret M Douglass, Li-Wei H Lehman, Andrew Reisner, Mauricio Villaroel, William J Long, Peter Szolovits, George B Moody, Roger G Mark, and Gari D Clifford. 2008. Automated de-identification of free-text medical records. *BMC Med Inform Decis Mak*, 8(32).

Mohammed Saeed, Christine Lieu, Greg Raber, and Roger G Mark. 2002. MIMIC II: a massive temporal ICU patient database to support research in intelligent patient monitoring. *Comput Cardiol*, 29:641–4.

Mohammed Saeed, Mauricio Villaroel, Andrew T. Reisner, Gari Clifford, Li-Wei Lehman, George Moody, Thomas Heldt, Tin H. Kyaw, Benjamin Moody, and Roger G. Mark. 2011. Multiparameter intelligent monitoring in intensive care ii (MIMIC-II): A public-access intensive care unit database. *Crit Care Med*, 39(5):952–60.

Pontus Stenetorp, Sampo Pyysalo, Goran Topić, Tomoko Ohta, Sophia Ananiadou, and Juni'chi Tsujii. 2012. BRAT: a web-based tool for NLP-assisted text annotation. In *Proc of EACL Demonstrations*, pages 102–7, Avignon, France. ACL.

Latanya Sweeney. 1996. Replacing personally-identifying information in medical records, the scrub system. In *AMIA Annu Fall Symp Proc*, pages 333–7, Washington, DC.

Özlem Uzuner, Yuan Luo, and Peter Szolovits. 2007. Evaluating the state-of-the-art in automatic de-identification. *J Am Med Inform Assoc*, 14(5):550–63.

An Analysis of Biomedical Tokenization: Problems and Strategies

Noa P. Cruz Díaz
University of Huelva, Huelva, Spain
noa.cruz@dti.uhu.es

Manuel M. Maña López
University of Huelva, Huelva, Spain
manuel.mana@dti.uhu.es

Abstract

Choosing the right tokenizer is a non-trivial task, especially in the biomedical domain, where it poses additional challenges, which if not resolved means the propagation of errors in successive Natural Language Processing analysis pipeline. This paper aims to identify these problematic cases and analyze the output that, a representative and widely used set of tokenizers, shows on them. This work will aid the decision making process of choosing the right strategy according to the downstream application. In addition, it will help developers to create accurate tokenization tools or improve the existing ones. A total of 14 problematic cases were described, showing biomedical samples for each of them. The outputs of 12 tokenizers were provided and discussed in relation to the level of agreement among tools.

1 Introduction

Tokenization is considered the first step in Natural Language Processing (henceforth, NLP) and it is broadly defined as the segmentation of text into primary building blocks for subsequent analysis (Webster and Kit, 1992).

Tokenization may seem simple if we assume that all it involves is the recognition of a space as a word separator (Baeza-Yates and Ribeiro-Neto, 2011). However, a closer examination will make it clear that a blank space alone is not enough even for general English (Jurafsky and Martin, 2009). Furthermore, choosing the right tokenization strategy is a non-trivial task, especially in the biomedical domain where it poses additional challenges (He and Kayaalp, 2006) which if not resolved means the propagation of errors in successive NLP analysis pipeline. As a consequence, text mining modules, such as Named Entity Recognition, will inevitably suffer in terms of effectiveness (Tomanek et al., 2007).

Tokenization in biomedical literature is particularly difficult due to the fact that general English differ from biomedical text in vocabulary and grammar (Barrett, 2012). In addition, scientific information has a particular structure (Harris, 2002). For example, Campbell and Johnson (2001) carried out three experiments to evaluate the syntactic dissimilarities between medical discharge summaries and everyday English, showing significant differences in syntactic content and complexity.

Another feature of the biomedical literature is related to terminology, which is inconsistently spelt and may vary from typographical errors to lower case and capitalized medication names (Krauthammer and Nenadic, 2004). Furthermore, biomedical texts could be ungrammatical (especially, clinical documents) as well as often include abbreviations and acronyms. Biomedical terms contain digits, capitalized letters within words, Latin and Greek letters, Roman digits, measurement units, list and enumerations, tabular data, hyphens and other special symbols. In addition, another complexity is the ambiguity, i.e., words and abbreviations that have different meanings (homonymy) and concepts described in more than one way (synonymy). For these reasons, the identification of terminology in the biomedical literature is one of the most challenging research topics in the last few years in NLP and biomedical communities and tokenization plays an important role in handling them.

There is no widely accepted tokenization method for English text, including biomedical documents since tokenization strategies can vary depending on language, task goals and other criteria. Previous approaches to biomedical tokenization lack guidance on how to modify existing tokenizers to new domains and how even to select them. Their idiosyncratic nature, detailed above, complicates this selection, modification and implementation (Barrett, 2012). Some authors also highlight the clear need for tokenization evaluation through the alignment and com-

Proceedings of the Sixth International Workshop on Health Text Mining and Information Analysis (Louhi), pages 40–49,
Lisbon, Portugal, 17 September 2015.

parison of the results of different tokenizers (Habert et al., 1998). To address this challenge, this paper identifies and describes all the problematic cases that can be found when tokenizing a biomedical text. In addition, it includes a list of useful tokenizers and a comparison of their outputs on biomedical text samples.

The rest of the paper is organized as follows. Firstly, the most relevant related research is outlined. Secondly, the tokenizers are listed and their outputs are shown. The paper finishes with conclusions.

2 Related Work

Despite its importance, tokenization is often neglected in the literature (Dridan and Oepen, 2012). Most research has been focused on annotating corpus with token information (Ohta, et al., 2002; Tanabe et al., 2005; Verspoor, et al., 2012) and developing or adapting tokenizers to new domains (Tomanek et al., 2007; McClosky and Charniak, 2008). However, little attention has been paid to the analysis of the problematic cases that appear in the tokenization process and the different strategies used for the current available tokenization tools to solve them.

To the best of our knowledge, for the biomedical domain, there is only one work devoted to a comparison of several tokenizers (He and Kayaalp, 2006). In this study, He and Kayaalp made a first approximation of the challenging cases. As authors affirmed, it can be considered as a starting point since the limited scope of their effort prevented them from developing a more complete set of cases. Especially, the instances identified for biomedical named entities are insufficient. The study also includes a comparison of the output of 13 tokenizers on 78 biomedical abstracts from Medline, a corpus of biomedical literature compiled by the U.S. National Library of Medicine.

Due to the limitations in the categorization of the complex cases and the fact that many tokenization tools have been developed in recent years, this paper complete all these cases, update the list of tokenization tools and test them on a set of biomedical sentences, outlining the differences among tokenization schemes. This means, providing a qualitative guideline for the reader which aid the decision making process of choosing the right tokenizer. This decision will depend mainly on the downstream task. In addition, the critical issues identified, allow developers to know what should be taken into account when adapting or developing tokenization tools.

3 Material and Methods

3.1 Problematic cases

We could divide the potential complexities in the tokenization process into two major categories: those that apply across all domains and those that are more likely to be found in biomedical corpora, where there is a large amount of technical vocabulary (Clegg, 2008). All these difficulties, together with sentences extracted from the BioScope corpus (Vincze et al., 2008), in which authors such as Velldal et al. (2012) found problematic cases where tokenizers fail, are detailed below:

Common English complexities

- **Hyphenated compound words**

For example:

(1) *Normal chest* ***x-ray****.*

(2) ***2-year 2-month*** *old female with pneumonia.*

(3) *This may occur through the ability of* ***IL-10*** *to induce expression of the gene.*

- **Words with letters and slashes**

Slashes usually indicate alternatives (e.g. *differentiation/activation*) or measurement units (e.g. *ng/ml*). In addition, they often separate two or more entity references (e.g. *IL-12/CD34*). They may also denote the knock-out status of a certain gene with respect to an organism (e.g. *flt3L-/-mice*) (Tomanek et al., 2007). For example:

(4) *The maximal effect is observed at the IL-10 concentration of 20* ***U/ml****.*

(5) *These results indicate that within the* ***TCR/CD3*** *signal transduction pathway both PKC and calcineurin are required for the effective activation of the IKK complex and NF-kappaB in T lymphocytes.*

- **Words with letters and apostrophes**

Apostrophes can indicate possessive (e.g. *years'*), words with single quotation (e.g. *'syntenic hits'*) and names (e.g. *O'Neill*). Examples of these might be the following:

(6) *The false positive rate of our predictor was estimated by the method of **D'Haeseleer** and Church 1855 and used to compare it to other prediction datasets.*

(7) *Small, scarred right kidney, below more than 2 standard deviations in size for **patient's** age.*

- **Words with letters and brackets**

There are basically four types of brackets: parentheses, square brackets, braces and angle brackets. For instance:

(8) *Of these, Diap1 has been most extensively characterized; it can block cell death caused by the ectopic expression of reaper, hid, and grim **(**reviewed in **[**26**])**.*

- **Abbreviations in capital letters and acronyms**

An abbreviation is a shortened form of a word or phrase. Usually, but not always, it consists of a letter or group of letters taken from the word or phrase. It must be taken into account in any tokenization process. An example of this may be the one shown below:

(9) *Mutants in Toll signaling pathway were obtained from **Dr.** S. Govind: cactE8, cactIIIG, and cactD13 mutations in the cact gene on Chromosome II.*

An acronym is an abbreviation formed from the initial components in a phrase or a word. These components may be individual letters (as in *SARS*; *severe acute respiratory syndrome*) or parts of words (as in *Ameslan*; *American Sign Language*).

Abbreviations and acronyms are commonly used in biomedical literature. For example, in the medical domain, writing favors brevity because time pressures often prevent medical specialists from describing clinical findings fully and abbreviations are a convenient way to shorten the sentences (Grange and Bloom, 2000).

Abbreviations and acronyms mainly refer to names, but abbreviations of adjectival expressions are often found in the biomedical domain (e.g. *CD8+* is an abbreviation of *CD8-positive*). For example:

(10) *The transcripts were detected in all the **CD4- CD8-**, **CD4+ CD8+**, **CD4+ CD8-**, and **CD4- CD8+** cell populations.*

- **Words with letters and periods**

Words with a period at the end usually indicate end of sentence. However, they may merely be abbreviations, such as *i.e.* and *e.g.* as shown in the following example:

(11) *Two stop codons of an iORF (**i.e.** the inframe and C-terminal stops) can be any combination of canonical stop codons (TAA, TAG, TGA).*

- **Words with letters and numbers**

For example:

(12) *Selenocysteine and pyrrolysine are the **21st** and **22nd** amino acids, which are genetically encoded by stop codons.*

- **Words with numbers and one type of punctuation**

Some simple examples for numbers are: large numbers (e.g. *390,926*), fractions (e.g. *1/2*), percentages (e.g. *50%*), decimals (e.g. *0.001*) and ranges (e.g. *2-5*). These punctuation marks are: comma, forward slash, percent, period and en dash. Good illustrations extracted from the BioScope corpus are the following:

(13) *A total of **26,003** iORF satisfied the above criteria.*

(14) *The patient had prior x-ray on **1/2** which demonstrated no pneumonia.*

(15) *Indeed, it has been estimated recently that the current yeast and human protein interaction maps are only **50%** and **10%** complete, respectively 18.*

(16) *The dotted line indicates significance level **0.05** after a correction for multiple testing.*

(17) *E-selectin is induced within **1–2** h, peaks at **4–6** h, and gradually returns to basal level by 24 h.*

- **Numeration**

It is regarded as the act or process of counting or numbering. For instance:

(18) ***1.*** *Bioactivation of sulphamethoxazole (SMX) to chemically-reactive metabolites and subsequent protein conjugation is thought to be involved in SMX hypersensitivity.*

- **A hypertext markup symbol**

Some of the frequently observed hypertext markup symbols are *<* and *"* (for the double quotation mark). For example:

(19) *Bcd mRNA transcripts of* ***<*** *or = 2.6 kb were selectively expressed in PBL and testis of healthy individuals.*

- **A URL**

An example would be the following:

(20) *Names of all available Trace Databases were taken from a list of databases at* ***http://www.ncbi.nlm.nih.gov/blast/mmtrace.shtml***

Biomedical English complexities

- **A DNA sequence**

For example:

(21) *Footprinting analysis revealed that the identical sequence* ***CCGAAACTGAAAAGG****, designated E6, was protected by nuclear extracts from B cells, T cells, or HeLa cells.*

- **Temporal expressions**

For instance:

(22) *This was last documented on the Nuclear Cystogram dated* ***1/2/01****.*

- **Chemical substances**

They include several symbols which may (or may not) denote word token boundary symbols such as parentheses, hyphens and slashes (Tomanek et al., 2007). Furthermore, chemical substances basically comprehend gene symbols, drug names and protein names, each of which has certain characteristics as described below.

<u>Gene symbols</u>

The names can indeed be divided into the following three categories (Proux et al., 1998).

- Names including special characters, i.e. upper cases, hyphen, digit, slash or brackets. For example, *Lam-B1* or *M(2)201*.
- Names in lower case and belonging to the general English language. For instance, *vamp* or *zip*.
- Names using lower case letters only without belonging to the language such as *zhr* or *sth*.

<u>Drug names</u>

In general, most drug names include:

- Particular letters from the chemical formula (e.g. T*ylenol*, which were generated from *n-aceryl-para-aminophenol*) as describe Gantner et al. (2002).
- Generic names such as *Thalomid.*
- Latin or Greek terminology.
- Parts or abbreviations of the company's name (e.g. *Baycol*, (*Bayer+colesterol*)).
- Low-frequency letters of the alphabet such as x or y (e.g. *x-trozine*).
- Acronyms like *Tigan* (that means *this is good against nausea*).

<u>Protein names</u>

Protein names can also be partitioned into three categories from their structure (Fukuda et al., 1998):

- Single words in upper case, numerical figures, and non-alphabetical letters which are mostly derived from gene name (e.g. *p53*).
- Compound words with upper case letters, numerical letters, and non-alphabetical letters. (e.g. *(IL-1)-responsive kinase*).
- Single word with only lower case letters (e.g. *insulin*).

Examples which appear in the BioScope corpus are the following:

(23) *These results reveal a central role for* ***CaMKIV/Gr*** *as a* ***Ca(2+)-regulated*** *activator of gene transcription in T lymphocytes.*

(24) *Expression of a highly specific protein inhibitor for cyclic AMP-dependent protein kinases in interleukin-1* ***(IL-1)-responsive*** *cells blocked* ***IL-1-induced*** *gene transcription that was driven by the kappa immunoglobulin enhancer or the human immunodeficiency virus long terminal repeat.*

3.2 Tokenization strategies

The tools analyzed were the following: Freeling, Genia tagger, Gate Unicode tokenizer (GUT), JULIE LAB tokenizer (JLT), LingPipe, McClosky-Charniak parser (MCP), MedPost, NLTK tokenizer, OpenNLP tokenizer, Penn Bio tokenizer, Stanford POS tagger and Xerox tokenizer. Table 1 details all these tokenizers showing their references and websites.

These tools were tested on the set of examples extracted from the BioScope corpus listed in the previous section. Tables 2 to 24 detail the output from each tokenizer. Each row of the tables shows the list of tokenizers with the same output. The numbers of the tools refer to Table 1. In bold, decisions in which tokenizers do not match.

The outputs, for which there is no agreement among several tools and, therefore, correspond to a single tokenizer, are not shown in this paper due to the space limit. However, this information can be found in Supplementary Material.

Common English complexities

- **Hyphenated compound words**

Table 2: Tokenizers output for sentence (1)

Tokenizer	Output
1, 2, 3, 6, 8, 9, 10, 11	Normal^chest^**x-ray**^.

Table 3: Tokenizers output for sentence (2)

Tokenizer	Output
1, 2, 6, 8, 9, 11, 12	**2-year^2-month**^old^female^with^pneumonia^.
3, 4, 5, 7	**2^-^year^2^-^month**^old^female^with^pneumonia^.

Table 4: Tokenizers output for sentence (3)

Tokenizer	Output
1, 2, 4, 6, 8, 9, 10, 11, 12	This^may^occur^through^the^ability^of^**IL-10**^to^induce^expression^of^the^gene^.^
5, 7	This^may^occur^through^the^ability^of^**IL^-^10**^to^induce^expression^of^the^gene^.^

- **Words with letters and slashes**

Table 5: Tokenizers output for sentence (4)

Tokenizer	Output
2, 6, 8, 9, 11, 12	The^maximal^effect^is^observed^at^the^IL-10^concentration^of^20^**U/ml**^.
3, 5, 7	The^maximal^effect^is^observed^at^the^IL^-^10^concentration^of^20^**U^/^ml**^.

Table 1: Overview of the 12 tools reviewed in the current study with their publications and website

	Tool	References	Website
1	Freeling	(Carreras, 2004; Padró and Stanilovsky, 2012)	http://nlp.lsi.upc.edu/freeling/
2	Genia	(Kulick et al., 2004; Tsuruoka et al., 2005; Tsuruoka and Tsujii, 2005)	http://www.nactem.ac.uk/tsujii/GENIA/tagger/
3	GUT	(Cunningham et al., 2002)	http://gate.ac.uk/sale/tao/splitch6.html#sec:annie:tokeniser
4	JLT	(Tomanek et al., 2007)	http://www.julielab.de/Resources/NLP+Tools.html
5	LingPipe	(Carpenter and Baldwin, 2011)	http://alias-i.com/lingpipe/
6	MCP	(McClosky and Charniak, 2008; McClosky, 2010)	http://nlp.stanford.edu/~mcclosky/biomedical.html
7	MedPost	(Smith et al., 2004)	ftp://ftp.ncbi.nlm.nih.gov/pub/lsmith/MedPost/medpost.tar.gz
8	NLTK	(Bird et al., 2009)	http://nltk.org/
9	OpenNLP	-	http://opennlp.apache.org/
10	Penn Bio	(Jin et al., 2006; McDonald and Pereira, 2005; McDonald et al., 2004)	http://www.seas.upenn.edu/~strctlrn/BioTagger/BioTagger.html
11	Stanford	(Toutanova et al., 2003)	http://nlp.stanford.edu/software/tagger.shtml
12	Xerox	(Beesley and Karttunen, 2003)	http://open.xerox.com/Services/fst-nlp-tools/Consume/175

1, 4, 10	The^maximal^effect^is^observed^ at^the^IL-10^concentration^of^20^ **U^/^ml.**

Table 6: Tokenizers output for sentence (5)

Tokenizer	Output
1, 2, 6, 8, 9, 11, 12	These^results^indicate^that^within^ the^**TCR/CD3**^signal^transduction^ pathway^both^PKC^and^calcineurin^ are^required^for^the^effective^activa tion^of^the^IKK^complex^and^ NF-kappaB^in^T^lymphocytes^.
3, 4, 5, 7, 10	These^results^indicate^that^within^ the^**TCR^/^CD3**^signal^transduction ^pathway^both^PKC^and^ calcineurin^are^required^for^the^ effective^activation^of^the^IKK^ complex^and^NF^-^kappaB^in^T^ lymphocytes^.

- **Words with letters and apostrophes**

Table 7: Tokenizers output for sentence (6)

Tokenizer	Output
1, 2, 4, 8, 9, 10, 11, 12	The^false^positive^rate^of^our^ predictor^was^estimated^by^the^ method^of^**D'Haeseleer**^and^ Church^1855^and^used^to^compare ^it^to^other^prediction^datasets^.
3, 5, 6, 7	The^false^positive^rate^of^our^ predictor^was^estimated^by^the^ method^of^**D^'^Haeseleer**^and^ Church^1855^and^used^to^compare ^it^to^other^prediction^datasets^.

Table 8: Tokenizers output for sentence (7)

Tokenizer	Output
1, 2, 4, 6, 8, 9, 10, 11, 12	Small^,^scarred^right^kidney^,^ below^more^than^2^standard^ deviations^in^size^for^**patient^'s^ age**^.
3, 5, 7	Small^,^scarred^right^kidney^,^ below^more^than^2^standard^ deviations^in^size^for^**patient^'^s**^ age^.

- **Words with letters and brackets**

Table 9: Tokenizers output for sentence (8)

Tokenizer	Output
1, 2, 5, 7, 8, 11, 12	Of^these^,^Diap1^has^been^most^ extensively^characterized^;^it^can^ block^cell^death^caused^by^the^ ectopic^expression^of^reaper^,^hid^ ,^and^grim^(^reviewed^in^**[^26^]^)**^.

- **Abbreviations in capital letters and acronyms**

Table 10: Tokenizers output for sentence (9)

Tokenizer	Output
4, 6, 8, 11	Mutants^in^Toll^signaling^pathway^ were^obtained^from^**Dr.**^S.^ Govind^:^cactE8^,^cactIIIG^,^and^ cactD13^mutations^in^the^cact^ gene^on^Chromosome^II^.
2, 5, 7	Mutants^in^Toll^signaling^pathway^ were^obtained^from^**Dr^.^S^.**^ Govind^:^cactE8^,^cactIIIG^,^and^ cactD13^mutations^in^the^cact^ gene^on^Chromosome^II^.

Table 11: Tokenizers output for sentence (10)

Tokenizer	Output
2, 6, 8, 9, 12	The^transcripts^were^detected^in^all ^the^**CD4-^CD8-**^,^**CD4+^CD8+**^ ,^**CD4+^CD8-**^,^and^**CD4-^CD8+**^ cell^populations^.
1, 3, 4, 7, 10, 11	The^transcripts^were^detected^in^all ^the^**CD4^-^CD8^-**^,^**CD4^+^ CD8^+**^,^**CD4^+^CD8^-**^,^and^ **CD4^-^CD8^+**^cell^populations^.

- **Words with letters and periods**

Table 12: Tokenizers output for sentence (11)

Tokenizer	Output
1, 6, 11, 12	Two^stop^codons^of^an^iORF^(^**i.e.**^the^inframe^and^C-terminal^ stops^)^can^be^any^combination^ of^canonical^stop^codons^(^TAA^ ,^TAG^,^TGA^)^.
2, 8	Two^stop^codons^of^an^iORF^(^ **i.e^.**^the^inframe^and^C-terminal^ stops^)^can^be^any^combination^ of^canonical^stop^codons^(^TAA^ ,^TAG^,^TGA^)^.
4, 7	Two^stop^codons^of^an^iORF^(^ **i^.^e^.**^the^inframe^and^**C^-^ terminal**^stops^)^can^be^any^ combination^of^canonical^stop^ codons^(^TAA^,^TAG^,^TGA^)^.

- **Words with letters and numbers**

Table 13: Tokenizers output for sentence (12)

Tokenizer	Output
1, 2, 4, 5, 6, 7, 8, 9, 11, 12	Selenocysteine^and^pyrrolysine^are^ the^**21st**^and^**22nd**^amino^acids^,^ which^are^genetically^encoded^by^ stop^ codons^.

- **Words with numbers and one type of punctuation**

Table 14: Tokenizers output for sentence (13)

Tokenizer	Output
1, 5, 6, 8, 9, 10, 11, 12	A˄total˄of˄**26,003**˄iORF˄satisfied˄ the˄above˄criteria˄.
2, 3, 4, 7	A˄total˄of˄**26˄,˄003**˄iORF˄satisfied ˄the˄above˄criteria˄.

Table 15: Tokenizers output for sentence (14)

Tokenizer	Output
1, 2, 6, 8, 9, 11, 12	The˄patient˄had˄prior˄**x-ray**˄on˄ **1/2**˄which˄demonstrated˄no˄ pneumonia˄.
4, 5, 7	The˄patient˄had˄prior˄**x˄-˄ray**˄ on˄**1˄/˄2**˄which˄demonstrated˄no˄ pneumonia˄.
3, 10	The˄patient˄had˄prior˄**x-ray**˄on˄ **1˄/˄2**˄which˄demonstrated˄no˄ pneumonia˄.

Table 16: Tokenizers output for sentence (15)

Tokenizer	Output
3, 4, 5, 6, 7, 8, 9, 10, 11	Indeed˄,˄it˄has˄been˄estimated˄ recently˄that˄the˄current˄yeast˄and˄ human˄protein˄interaction˄maps˄ are˄only˄**50˄%**˄and˄**10˄%**˄ complete˄,˄respectively˄18˄.

Table 17: Tokenizers output for sentence (16)

Tokenizer	Output
1, 2, 4, 5, 6, 8, 9, 10, 11, 12	The˄dotted˄line˄indicates˄ significance˄level˄**0.05**˄after˄a˄ correction˄for˄multiple˄ testing˄.
3, 7	The˄dotted˄line˄indicates˄ significance˄level˄**0˄.˄05**˄after˄a˄ correction˄for˄multiple˄ testing˄.

Table 18: Tokenizers output for sentence (17)

Tokenizer	Output
1, 2, 8, 9, 10, 11, 12	E-selectin˄is˄induced˄within˄ **1–2**˄**h**˄,˄peaks˄at˄**4–6**˄**h**˄,˄and˄ gradually˄returns˄to˄basal˄level˄by˄ **24**˄**h**˄.
4, 7	E-selectin˄is˄induced˄within˄ **1˄–˄2˄h**˄,˄peaks˄at˄**4˄–˄6˄h**˄,˄ and˄gradually˄returns˄to˄basal˄ level˄by˄ **24˄h**˄.

- **Numeration**

Table 19: Tokenizers output for sentence (18)

Tokenizer	Output
1, 2, 3, 5, 7, 8, 9, 10, 11, 12	**1**˄.˄Bioactivation˄of˄sulphamethoxaz ole˄(˄SMX˄)˄to˄ chemically-reactive˄metabolites˄and˄ subsequent˄protein˄conjugation˄is˄ thought˄to˄be˄involved˄in˄SMX˄ hypersensitivity˄.
4, 6	**1.**˄Bioactivation˄of˄ sulphamethoxazole˄(˄SMX˄)˄to˄ chemically-reactive˄metabolites˄ and˄subsequent˄protein˄conjugation˄ is˄thought˄to˄be˄involved˄in˄ SMX˄hypersensitivity˄.

- **A hypertext markup symbol**

Table 20: Tokenizers output for sentence (19)

Tokenizer	Output
2, 4, 5, 8	Bcd˄mRNA˄transcripts˄of˄**&˄lt˄;**˄ or˄=˄**2.6**˄**kb**˄were˄selectively˄ expressed˄in˄PBL˄and˄testis˄of˄ healthy˄individuals˄.
9, 12	Bcd˄mRNA˄transcripts˄of˄**&lt˄;**˄ or˄=˄**2.6**˄**kb**˄were˄selectively˄ expressed˄in˄PBL˄and˄testis˄of˄ healthy˄individuals˄.
3, 7	Bcd˄mRNA˄transcripts˄of˄**&˄lt˄;**˄ or˄=˄**2˄.˄6**˄**kb**˄were˄selectively˄ expressed˄in˄PBL˄and˄testis˄of˄ healthy˄individuals˄.

- **A URL**

Table 21: Tokenizers output for sentence (20)

Tokenizer	Output
2, 6, 8	Names˄of˄all˄available˄Trace˄ Databases˄were˄taken˄from˄a˄list˄ of˄databases˄at˄**http˄:˄//www.ncbi. nlm.nih.gov/blast/mmtrace.shtml**
3, 5, 7	Names˄of˄all˄available˄Trace˄ Databases˄were˄taken˄from˄a˄list˄ of˄databases˄at˄**http˄:˄/˄/˄www˄ .˄ncbi˄.˄nlm˄.˄nih˄.˄gov˄/˄blast˄ /˄mmtrace˄.˄shtml**
11, 12	Names˄of˄all˄available˄Trace˄ Databases˄were˄taken˄from˄a˄list˄ of˄databases˄at˄**http://www.ncbi. nlm.nih.gov/blast/mmtrace.shtml**

Biomedical English complexities

- **A DNA sequence**

Table 22: Tokenizers output for sentence (21)

Tokenizer	Output

1, 2, 4, 5, 6, 7, 8, 9, 11, 12	Footprinting‸analysis‸revealed‸that‸the‸identical‸sequence‸**CCGAAACT GAAAAGG**‸,‸designated‸E6‸,‸was‸protected‸by‸nuclear‸extracts‸from‸B‸cells‸,‸T‸cells‸,‸or‸HeLa‸cells‸.

- **Temporal expressions**

Table 23: Tokenizers output for sentence (22)

Tokenizer	Output
2, 6, 8, 9, 11, 12	This‸was‸last‸documented‸on‸the‸Nuclearv‸Cystogram‸dated‸**1/2/01**‸.
1, 3, 4, 7, 10	This‸was‸last‸documented‸on‸the‸Nuclearv‸Cystogram‸dated‸**1**‸**/2**‸**/**‸**0 1**‸.

- **Chemical substances**

Table 24: Tokenizers output for sentence (23)

Tokenizer	Output
6, 8	These‸results‸reveal‸a‸central‸role‸for‸**CaMKIV/Gr**‸as‸a‸**Ca**‸**(2+**‸**)**‸**-regulated**‸activator‸of‸gene‸transcription‸in‸T‸lymphocytes‸.
1, 3, 4, 7	These‸results‸reveal‸a‸central‸role‸for‸**CaMKIV**‸**/**‸**Gr**‸as‸a‸**Ca**‸**(**‸**2+**‸**)**‸**-**‸**regulated**‸activator‸of‸gene‸transcription‸in‸T‸lymphocytes‸.

Table 25: Tokenizers output for sentence (24)

Tokenizer	Output
1, 2, 6, 8, 11	Expression‸of‸a‸highly‸specific‸protein‸inhibitor‸for‸cyclic‸**AMP-dependent**‸protein‸kinases‸in‸**interleukin-1**‸**(**‸**IL-1**‸**)**‸**-**‸**responsive**‸cells‸blocked‸**IL-1-induced**‸gene‸transcription‸that‸was‸driven‸by‸the‸kappa‸immunoglobulin‸enhancer‸or‸the‸human‸immunodeficiency‸virus‸long‸terminal‸repeat‸.

4 Conclusions

This paper analyzed the problematic cases that can be found when tokenizing a biomedical text. In addition, it listed a set of potentially useful tokenizers and tested them on biomedical sentences.

Identifying the complex cases that introduce this domain and knowing what types of behavior are expected from available tokenizers in each of these cases is vital. This will enable researchers to be aware of those aspects which are especially challenging when developing new tools or adapting existing ones. In addition, it will aid the process of selecting the right tokenizer according to the most appropriate tokenization scheme for the downstream application. This will facilitate to lose the minimum of information. Obviously, other factors like technical, usability of functional criteria should be taken into account in such decision.

The experiments carried out showed a widely variation on the results. This variability was expected since there is no a single tokenization method. Neither of the tools produced identical output. Tokenizers pair that coincided in the same strategy or scheme in over 75% of cases were Genia tagger and NLTK tokenizer as well as Stanford POS tagger and NLTK tokenizer.

Regarding the challenging problems where there was more disagreement (less than 35% agreement) and, therefore, presented more difficulties for the tokenization tools are, the hypertext markup symbol, URLs and chemical substances. The latter was assumed since biomedical terminology is currently one of the most challenging research topics in NLP.

Among the cases with more than 80% agreement, it can be found: hyphenated compound words, words with letters and numbers, words with numbers and one type of punctuation and DNA sequences.

References

Andrew B. Clegg. 2008. Computational-Linguistic Approaches to Biological Text Mining. PhD thesis. School of Crystallography Birkbeck, University of London.

Benoit Habert, Gilles Adda, M. Adda-Decker, P. Boula de Marëuil, S. Ferrari, O. Ferret, G. Illouz, and P. Paroubek. 1998. Towards tokenization evaluation. In *Proceedings of the 1st International Language Resources and Evaluation*, p. 427-431. Granada, Spain.

Bob Carpenter, and Breck Baldwin. 2011. Natural Language Processing with LingPipe 4. LingPipe Publishing, New York.

Bob Grange, D.A. Bloom. 2000. Acronyms, abbreviations and initialisms. *BJU international*, vol. 86, no 1, p. 1-6.

Dan Jurafsky and James H. Martin. 2009. *Speech and Language Processing, 2nd Edition*. NJ: Pearson.

David Campbell and Stephen Johnson. 2001. Comparing syntactic complexity in medical and non-medical corpora. In *Proceedings of the AMIA Symposium,* p. 90. American Medical Informatics Association.

David McClosky, Eugene Charniak. 2008. Self-Training for Biomedical Parsing. In *Proceedings of the Association for Computational Linguistics*. Columbus, Ohio.

David McClosky. 2010. Any Domain Parsing: Automatic Domain Adaptation for Natural Language Parsing. PhD thesis. Department of Computer Science, Brown University.

Denys Proux, Francois Rechenmann, Laurent Julliard, Violaine Pillet, and Bernard Jacq. 1998. Detecting gene symbols and names in biological texts: a first step toward pertinent information extraction. *Genome informatics series*, 72-80.

Erik Velldal, Lilja Øvrelid, Jonathon Read and Stephan Oepen. 2012. Speculation and negation: Rules, rankers, and the role of syntax. *Computational linguistics*, 38:2, 369-410.

Florian Gantner, Christian Schweiger, and Michael Schlander. 2002. Naming, classification, and trademark selection: implications for market success of pharmaceutical products. *Drug information journal*, 36:807–824.

Hamish Cunningham, Diana Maynard, Kalina Bontcheva, Valentin Tablan. 2002. GATE: an architecture for development of robust HLT applications. In *Proceedings of the 40th Annual Meeting on Association for Computational Linguistics*, p. 168-175. Portland, OR, USA.

Jonathan Webster and Chunyu Kit. 1992. Tokenization as the initial phase in NLP. In *Proceedings of the 14th Conference on Computational Linguistics-Volume 4,* 1106-1110.

Katrin Tomanek, Joachim Wermter, and Udo Hahn. 2007. Sentence and token splitting based on conditional random fields. In *Proceedings of the 10th Conference of the Pacific Association for Computational Linguistics*, p. 49-57. Portland, OR, USA.

Karin Verspoor, et al. 2012. A corpus of full-text journal articles is a robust evaluation tool for revealing differences in performance of biomedical natural language processing tools. *BMC bioinformatics* 13.1 (2012): 207.

Kenneth R. Beesley and Lauri Karttunen. 2003. Finite state morphology. *Stanford: CSLI publications.*

K. I. Fukuda, T. Tsunoda, A. Tamura, and T Takagi. 1998. Toward information extraction: identifying protein names from biological papers. *Pac. Symp. Biocomput*, 707-18.

Kristina Toutanova, Dan Klein, Chritopher D. Manning, and Yoram Singer. 2003. Feature-rich part-of-speech tagging with a cyclic dependency network. In *Proceedings of the 2003 Conference of the North American Chapter of the Association for Computational Linguistics on Human Language Technology*, p. 173-180. Portland, OR, USA.

Lorraine Tanabe, Natalie Xie, Lynne H Thom, Wayne Matten, and W John Wilbur. 2005. GENETAG: a tagged corpus for gene/protein named entity recognition. *BMC Bioinformatics*, 6(suppl 1):S3.

Lluís Padró, and Evgeny Stanilovsky. 2012. Freeling 3.0: towards wider multilinguality. In *Proceedings of the 8th International Language Resources and Evaluation*. Istanbul, Turkey.

L. Smith, T. Rindflesch, and W.J. Wilbur. 2004. MedPost: a part-of-speech tagger for bioMedical text. Bioinformatics, 20:14, 2320-2321.

Michael Krauthammer and Goran Nenadic. 2004. Term identification in the biomedical literature. *Journal of biomedical informatics*, 37:6, 512-526.

Neil Barrett. 2012. Natural language processing techniques for the purpose of sentinel event information extraction. PhD thesis. University of Victoria, Canada.

Rebecca Dridan and Stephan Oepen. 2012. Tokenization: Returning to a Long Solved Problem. In *Proceedings of the 50th Annual Meeting of the Association for Computational Linguistics*, p.

378-382. Association for Computational Linguistics, Jeju, Korea.

Ricardo Baeza-Yates and Berthier Ribeiro-Neto. 2011. *Modern Information Retrieval - the concepts and technology behind search, Second edition*. Pearson Education Ltd., Harlow, England.

Ryan McDonald, and Fernando Pereira. 2005. Identifying gene and protein mentions in text using conditional random fields. *BMC bioinformatics*, 6:S1, S6.

Ryan T. McDonald, R. Scott Winters, Mark Mandel, Yang Jin, Peter S. White, and Fernando Pereira. 2004. An entity tagger for recognizing acquired genomic variations in cancer literature. *Bioinformatics*, 20:17, 3249-3251.

Seth Kulick, Ann Bies, Mark Liberman, Mark Mandel, Ryan McDonald, Martha Palmer, Andrew Schein, and Lyle Ungar. 2004. Integrated Annotation for Biomedical Information Extraction. *NAACL/HLT Workshop on Linking Biological Literature, Ontologies and Databases: Tools for Users*, 61-68. Boston.

Steven Bird, Ewan Klein, Edward Loper. 2009. *Natural language processing with Python*. O'Reilly Media.

Tomoko Ohta, Yuka Tateisi, Jin-Dong Kim. 2002. GENIA corpus: an annotated research abstract corpus in molecular biology domain. In *Proceedings of the Human Language Technology Conference*, 73-77. San Diego, California.

Veronika Vincze, György Szarvas, Richárd Farkas, György Móra, and János Csirik. 2008. The BioScope corpus: biomedical texts annotated for uncertainty, negation and their scopes. *BMC bioinformatics*, 9(Suppl 11), S9.

Xavier Carreras, Isaac Chao, Lluís Padró, and Muntsa Padró. 2004. Freeling: An open-source suite of language analyzers. In *Proceedings of the 4th International Language Resources and Evaluation*,. Lisbon, Portugal.

Yang Jin, Ryan T McDonald, Kevin Lerman, Mark A. Mandel, Steven Carroll, Mark Y. Liberman, Fernando C. Pereira, Raymond S. Winters, and Peter S White. 2006. Automated recognition of malignancy mentions in biomedical literature. *BMC bioinformatics*, 7:1, 492.

Ying He and Mehmet Kayaalp. 2006. A Comparison of 13 Tokenizers on MEDLINE. The Lister Hill National Center for Biomedical Communications, Tech. Rep. LHNCBC-TR-2006-003.

Yoshimasa Tsuruoka, Yuka Tateishi, Jin-Dong Kim, Tomoko Ohta, John McNaught, Sophia Ananiadou, Jun'ichi Tsujii. 2005. Developing a robust part-of-speech tagger for biomedical text. *Advances in informatics - 10th Panhellenic Conference on Informatics*. Springer Berlin Heidelberg, p. 382-392.

Yoshimasa Tsuruoka, and Jun'ichi Tsujii. 2005. Bidirectional inference with the easiest-first strategy for tagging sequence data. In *Proceedings of the conference on Human Language Technology and Empirical Methods in Natural Language Processing*, p. 467-474. Portland, OR, USA.

Zellig S. Harris. 2002. The structure of science information. *Journal of biomedical informatics*, 35:4, 215-221.

Annotation of Clinically Important Follow-up Recommendations in Radiology Reports

Meliha Yetisgen[1,2], Prescott Klassen[2], Lucas H. McCarthy[3], Elena Pellicer[4], Thomas H. Payne[4,5], Martin L. Gunn[6]
Department of Biomedical Informatics and Medical Education[1], Department of Linguistics[2], Department of Neurology[3], School of Medicine[4], Information Technology Services[5], Department of Radiology[6]
University of Washington
Seattle, WA
melihay,klassp,lucasmc,pellicer,tpayne,marting@uw.edu

Abstract

Communication of follow-up recommendations when abnormalities are identified on imaging studies is prone to error. The absence of an automated system to identify and track radiology recommendations is an important barrier to ensuring timely follow-up of patients especially with non-acute incidental findings on imaging studies. We are in the process of building a natural language processing (NLP) system to identify follow-up recommendations in free-text radiology reports. In this paper, we describe our efforts in creating a multi-institutional radiology report corpus annotated for follow-up recommendation information. The annotated corpus will be used to train and test the NLP system.

1 Introduction

A radiology report is the principal means by which radiologists communicate the findings of an examination to the referring physician and sometimes the patient. With the dramatic rise in utilization of medical imaging in the past two decades, health providers are challenged by the optimal use of clinical information while not being overwhelmed by it. Based on potentially important observations the radiologist may recommend specific imaging tests or a clinical follow-up in the narrative radiology report. These recommendations are made for several potential reasons. The radiologist may recommend further investigation to clarify the diagnosis or exclude potentially serious, but clinically expected disease. Secondly, the radiologist may unexpectedly encounter signs of potentially serious disease on the imaging study that they believe require further investigation. Thirdly, the radiologist may recommend surveillance of disease to ensure an indolent course. Finally, a radiologist may provide advice to the referring physician about the most effective future test(s) specific to the patient's disease or risk factors.

The reliance on human communication, documentation, and manual follow-up is a critical barrier to ensuring that appropriate imaging or clinical follow-up occurs. The World Alliance of Patient Safety, a part of the World Health Organization, recently identified poor test results follow-up as one of the major processes contributing to unsafe patient care[1].

There are many potential failure points when communicating and following up on important radiologic findings and recommendations: (1) Critical findings and follow-up recommendations not explicitly highlighted by radiologists: Although radiologists describe important incidental observations in reports, they may or may not phone an ordering physician. If these recommendations "fall through the cracks" patients may present months later with advanced disease (e.g., metastatic cancer). (2) Patient mobility: When patients move between services in healthcare facilities, there is increased risk during "hand-offs" of problems with test result follow-up and continuity of care (Callen et al., 2011). (3) Heavy workload of providers: Physicians and other pro-

[1] World Alliance for Patient Safety. Summary of the Evidence on Patient Safety: Implications for Research. Geneva: World Health Organization, 2008. Accessed: 3.13.2015. Available at: http://gawande.com/documents/WHOGuidelinesforSafeSurgery.pdf

Proceedings of the Sixth International Workshop on Health Text Mining and Information Analysis (Louhi), pages 50–54,
Lisbon, Portugal, 17 September 2015.

viders have to deal with a deluge of test results. A survey of 262 physicians at 15 internal medicine practices found that physicians spend on average 74 minutes per clinical day managing test results, and 83% of physicians reported at least one delay in reviewing test results in the previous two months (Holden et al., 2004). However, it is vital that these results, particularly if they are unexpected, are not lost to follow-up. In patients who have an unexpected finding on a chest radiograph, approximately 16% will eventually be diagnosed with a malignant neoplasm (Poon et al., 2004).

These examples indicate an opportunity to develop a systematic approach to augmenting existing channels of clinical information for preventing delays in diagnosis. The goals of our research are to: (1) define clinically important recommendations in the context of radiology reports and (2) create a large-scale radiology report corpus annotated with recommendation information. The corpus will be used to build an automated system that will extract recommendation information so that reports can be flagged visually and electronically.

2 Related Work

Identifying follow-up recommendation information in radiology reports has been previously studied by other researchers. Dreyer et al. processed 1059 radiology reports with Lexicon Mediated Entropy Reduction (LEXIMER) to identify the reports that include clinically important findings and recommendations for subsequent action (Dreyer et al., 2005). The same research group performed a similar analysis on a database of radiology reports covering the years 1995-2004 (Dang et al., 2008). From that database, they randomly selected 120 reports with and without recommendations. Two radiologists independently classified those selected reports according to the presence of recommendation, time-frame, and imaging-technique suggested for follow-up examination. These reports were analyzed by an NLP system first for classification into two categories: reports with recommendations and reports without recommendations. The reports with recommendations were then classified into those with imaging recommendations and those with non-imaging recommendations. The recommended time frames were identified and normalized into a number of days. The authors reported 100% accuracy in identifying reports with and without recommendations. In 88 reports with recommendation, they reported 0.945 precision in identifying temporal phrases, and 0.932 in identifying recommended imaging tests. In a follow-up study, the authors analyzed the rate of recommendations by performing a statistical analysis on 5.9 million examinations (Sistrom et al., 2009). In all three papers, they reported high overall performance values; however, the authors presented their text processing approach as a black box without providing nec-

CT ABDOMEN AND PELVIS WITH INTRAVENOUS CONTRAST
HISTORY:
Prostate CA-Prostate CA Surveillance
COMPARISON: None
CONTRAST: iv contrast was used. Positive oral contrast was administrated
TECHNIQUE:
Region of interest: Abdomen-Pelvis
Superior Extent: Diaphragm. Inferior Extent: Symphysis Pubis
....
FINDINGS:
Lung bases: A 6-mm nodule is noted in the peripheral left lung base (image 9, series 2). There is a focal area of
atelectasis in the anterior right lung base.
Pleura: No pleural effusions or thickening.
Included heart: No gross abnormality..
Liver: Normal
Portal veins: Normal.
Gallbladder and bile ducts: Normal
Spleen: Normal
Aorta and IVC: There is atherosclerotic calcification of the aorta.
There is a small focus of ulcerated plaque in the infrarenal aorta (image 49, series 2).
Stomach, duodenum and small bowel: Normal
....
IMPRESSION:
1. Incidental 6-mm left lung nodule. Follow-up chest CT is recommended in 6 months.
2. A few prostatic calcifications are noted. No CT evidence of metastatic prostate cancer.
3. Small ulcerated atheromatous plaque in the infrarenal aorta.

Figure 1: Example radiology report with follow-up recommendation

essary information required to replicate their methods.

3 Follow-up Recommendations in Radiology Reports

In this research, we define a *follow-up recommendation* as a statement made by the radiologist in a given radiology report to advise the referring clinician to further evaluate an imaging finding by either other tests or further imaging. Figure 1 presents a radiology report with such a follow-up recommendation (Line 24: Incidental 6-mm left lung nodule. *Follow-up chest CT is recommended in 6 months*).

Under the supervision of a radiologist and an internal medicine specialist, we analyzed a small set of radiology reports with different modalities and grouped the follow-up recommendations under the following four non-overlapping categories.

Category 1: Non-contingent clinically important recommendation: An advisory statement that could result in mortality or significant morbidity if appropriate clinical assessment, diagnostic or therapeutic follow-up steps are not followed.
Case example: Incidental lung mass suspicious for malignancy on a trauma CT of the abdomen.
Follow-up recommendation example: *CT chest is recommended to further evaluate the lung mass.*

Category 2: Contingent clinically important recommendation: Similar to (a), but the statement is conditional on the presence of a clinical condition.
Case example: Adrenal mass identified on a CT of the abdomen and pelvis for appendicitis.
Follow-up recommendation example: *If the patient has a history of malignancy, consider biochemical testing and an adrenal mass protocol CT for further evaluation.*

Category 3: Clinically important recommendation likely reported: Similar to (a) and (b), but considered to be unlikely not to be reported in communication between radiologist and clinician.
Case example: A distal radius fracture was identified on a previous week's x-ray of patient's hand. A follow-up x-ray of the hand is requested to rule out possible additional scaphoid fracture.
Follow-up recommendation example: *L distal radius fracture x 1 week, please also follow-up to rule out scaphoid fracture compared with last week's x-rays.*

Category 4: Clinically unimportant recommendation: An advisory statement that is unlikely to result in mortality or significant morbidity if appropriate clinical assessment, diagnostic or therapeutic follow-up steps are not followed, and/or a low probability that the recommendation would be overlooked.
Case example: Following trauma, a radiograph demonstrates a probable non-displaced fracture of the mid ulna.
Follow-up recommendation example: *Consider an MRI of the forearm if diagnostic certainty is desired.*

To capture the main attributes of follow-up recommendations, we created a simple template with three entities; reason for recommendation (e.g., *incidental 6-mm left lung nodule*), recommended test (e.g., *chest CT*), and time-frame (e.g., in *6 months*). We use the follow-up recommendation categories and template to annotate a large scale radiology corpus that will be explained in the following sections.

4 Corpora for Follow-up Recommendations

4.1 Pilot Corpus

Dataset: In previous work, we created a corpus of radiology reports composed of 800 de-identified radiology reports extracted from the radiology information system of our institution (Yetisgen-Yildiz et al., 2013). The reports represented a mixture of imaging modalities, including radiography computer tomography (CT), ultrasound, and magnetic resonance imaging (MRI). The distribution of the reports across imaging modalities is listed in Table 1.

Imaging modality	Frequency
Computer tomography	486
Radiograph	259
Magnetic resonance imaging	45
Ultrasound	10
Total	800

Table 1: Distribution of reports in pilot corpus.

Annotation Guidelines: We annotated this dataset prior to defining different categories of follow-up recommendations. In this annotation task, we asked the annotators simply to highlight the boundaries of sentences that include any follow-up recommendation.

Annotation Process: Two annotators, one radiologist and one internal medicine specialist, went through each of the 800 reports and marked the sentences that contained follow-up recommendations. Out of 18,748 sentences in 800 reports, the radiologist annotated 118 sentences and the clinician annotated 114 sentences as recommendation. They agreed on 113 of the sentences annotated as recommendation. The inter-rater agreement was 0.974 F-score.

4.2 Multi-institutional Radiology Corpus

We extended our pilot dataset of 800 reports with a much larger set of 745,058 radiology reports from three different institutions including University of Washington Medical Center, Harborview Medical Center, and Seattle Cancer Care Alliance. The corpus covers the full range of imaging modalities, including radiographs, computed tomography, ultrasound, and magnetic resonance imaging (Table 2).

Imaging modality	Frequency
Computed Radiography	413,889
Computed Tomography	146,181
Digital Fluoroscopy	12
Digital Radiography	1,626
Magnetic Resonance Imaging	52,127
Nuclear Medicine	12,895
Portable Radiography	6,166
Portable Radiography	4,121
Fluoroscopy	27,239
Ultrasound	68,999
Angio-Interventional	11,803
Total	745,058

Table 2: Distribution of reports in multi-institutional radiology corpus.

We excluded the Mammography modality, which was comprised of 37,754 reports because a specific follow-up and alert system was already in place.

Annotation Guidelines: We designed the annotation task to operate on two levels; sentence level and entity level. At the sentence level, the annotators mark the boundaries of recommendation sentences and label each marked sentence with one of the four recommendation categories: (1) *non-contingent clinically important recommendation*, (2) *contingent clinically important recommendation*, (3) *clinically important recommendation likely reported*, and (4) *clinically unimportant recommendation*. At the entity level, the annotators mark the three attributes of recommendation information presented in the marked sentences: (1) *reason for follow-up recommendation*, (2) *recommended follow-up test*, and (3) *time-frame for follow-up test.*

Annotation Process: Because manual annotation is a time-consuming and labor-intensive process, we could annotate only a small portion of our large radiology corpus. The percentage of reports that include recommendation sentences is quite low—about 15% at our institution. To increase the number of reports with recommendations in the annotated set, rather than randomly sampling, we built a high recall (0.90), low precision (0.35) classifier trained on the pilot dataset described in section 4.1. The details of this baseline classifier can be found in our prior publication (Yetisgen-Yildiz et al., 2013). We ran our baseline classifier on un-annotated reports and only sampled reports for manual annotation from the reports our classifier identified as positive for follow-up recommendations. Because the classifier was high recall but low precision, it identified many false positives. The filtering of reports using a classifier reduced the number of reports our human annotators needed to review, expediting the annotation process.

At the sentence level, one radiologist and one neurologist review the classifier-selected reports with system generated follow-up recommendation sentences highlighted. The annotators correct the system generated sentences and/or highlight new sentences if needed. They associate each highlighted sentence with one of the four types described in Section 3.

At the entity level, one neurologist and one medical school student annotate the entities (reason for recommendation, recommended test, and time frame) in reports annotated in a previous stage at the sentence level with follow-up recommendations.

Inter-annotator Agreement Levels: At the sentence level, we measured the inter-annotator agreement on a set of 50 reports featuring at least one system-generated recommendation identified by our high recall classifier from a randomly selected collection of one thousand reports. Our annotation process required annotators to re-label all sentences that were initially identified by the system as a recommendation with the four type-specific labels described in Section 3. They could label the sentence as *Incorrect* if they believed the system had wrongly identified a recommendation sentence and they could also label a new recommendation sentence if they believed it had not been identified correctly by the system. The inter-rater agreement levels were kappa 0.43 and 0.59 F1 score. To resolve the disagreements, we scheduled multiple meetings. One of our observations during those meetings was that none of the new recommendation sentences introduced by either annotator were identified by the other. In our review, both annotators agreed that the majority of the new recommendations the other introduced were correct. We adjusted our annotation guidelines to add rules to help decide if and when a new sentence should be identified as a recommendation.

At the entity level, agreement levels were 0.78 F1 for reason, 0.88 F1 for test, and 0.84 F1 for time frame.

Our annotation process is on-going. The annotators completed the annotation of 567 radiology reports using updated guidelines based on the inter-annotator agreement stage. They highlighted 265 sentences as category 1, 90 sentences as category 2, 222 sentences as category 3, and 160 sentences as category 4. At the entity level, for 225 recommendation sentences, the annotators highlighted 207 text spans as reason, 314 text spans as test, and 71 text spans as time-frame.

5 Conclusion

In this paper, we described our efforts in creating a large scale radiology corpus annotated for follow-up recommendations. We are in the process of building a text processing system based on our current annotated corpus.

References

Callen J, Georgiou A, Li J, Westbrook JI. The safety implications of missed test results for hospitalized patients: a systematic review. BMJ Qual Saf. 2011;20(2):194-9.

Dang PA, Kalra MK, Blake MA, Schultz TJ, Halpern EF, Dreyer KJ. Extraction of Recommendation Features in Radiology with Natural Language Processing: Exploratory Study. AJR. 2008; 191:313-20.

Dreyer KJ, Kalra MK, Maher MM, Hurier AM, Asfaw BA, Schultz T, Halpern EF, Thrall JH. Application of Recently Developed Computer Algorithm for Automatic Classification of Unstructured Radiology Reports: Validation Study. Radiology. 2005; 234:323-39.

Holden WE, Lewinsohn DM, Osborne ML, Griffin C, Spencer A, Duncan C, Deffebach ME. Use of a clinical pathway to manage unsuspected radiographic findings. Chest. 2004;125(5):1753-60.

Poon EG, Gandhi TK, Sequist TD, Murff HJ, Karson AS, Bates DW. "I wish I had seen this test result earlier!": Dissatisfaction with test result management systems in primary care. Arch Intern Med. 2004;164(20):2223-8.

Sistrom CL, Dreyer KJ, Dang PP, Weilburg JB, Boland GW, Rosenthal DI, Thrall JH. Recommendations for Additional Imaging in Radiology Reports: Multifactorial Analysis of 5.9 Million Examinations. Radiology. 2009; 253(2):453-61.

Xia F, Yetisgen-Yildiz M. Clinical Corpus Annotation: Challenges and Strategies. Proceedings of Third Workshop on Building and Evaluating Resources for Biomedical Text Mining (BioTxtM'2012) of the International Conference on Language Resources and Evaluation (LREC), Istanbul, May, 2012.

Yetisgen-Yildiz M, Gunn ML, Xia F, Payne TH. A Text Processing Pipeline to Extract Recommendations from Radiology Reports. J Biomed Inform., 2013;46(2):354-362.

On the Impact of Twitter-based Health Campaigns: A Cross-Country Analysis of Movember

Nugroho Dwi Prasetyo &
Claudia Hauff
Web Information Systems
Delft University of Technology
Delft, the Netherlands

Dong Nguyen &
Tijs van den Broek &
Djoerd Hiemstra
University of Twente
Enschede, the Netherlands

Abstract

Health campaigns that aim to *raise awareness* and subsequently *raise funds* for research and treatment are commonplace. While many local campaigns exist, very few attract the attention of a global audience. One of those global campaigns is `Movember`, an annual campaign during the month of November, that is directed at men's health with special foci on cancer & mental health. Health campaigns routinely use social media portals to capture people's attention. Recently, researchers began to consider to what extent social media is effective in raising the awareness of health campaigns. In this paper we expand on those works by conducting an investigation across four different countries, while not only restricting ourselves to the impact on awareness but also on fund-raising. To that end, we analyze the 2013 `Movember` Twitter campaigns in Canada, Australia, the United Kingdom and the United States.

1 Introduction

The rise of social media portals — and thus access to vast amounts of user-generated data — has not gone unnoticed within the health care domain. Existing works have, amongst others, exploited social media data to track and predict the spread of diseases (Achrekar et al., 2011; Culotta, 2010; Chew and Eysenbach, 2010; Diaz-Aviles and Stewart, 2012), to analyse the effects of drug interactions (Segura-Bedmar et al., 2014), and to examine trends for cardiac arrest and resuscitation communication (Bosley et al., 2013).

Social media portals have also been employed to distribute health information on diseases and treatment options. In (Scanfeld et al., 2010; Vance et al., 2009), for instance, it has been shown that effective dissemination of such information can be achieved through Twitter and YouTube. At the same time though, Moorhead et al. (2013) argue that social health communication research is still in its infancy and large gaps in our understanding remain.

While the usage of social media for health campaigns is ever-growing, very few works have considered how *effective* these campaigns are in achieving their goals. While Thackeray et al. (2013) and Bravo and Hoffman-Goetz (2015) investigated the change of people's awareness during social media health campaigns, to our knowledge no research so far has considered the second goal of many health campaigns — raising funds for research and treatment.

In this paper, we contribute to closing this gap, (1) by conducting an awareness-based large-scale analysis *across several countries*, and (2) by investigating the extent to which a global social-media based health campaign is successful in terms of fund-raising. We investigate the particular use case of `Movember`, an annual health campaign conducted (amongst others) through social media channels that has two goals[1]: (1) to gather *"funding for the `Movember` Foundation's men's health programs"*, and, (2) to start *"conversations about men's health"*. In both cases, the main foci are on various types of cancer that typically occur in men and on men's mental health. `Movember` is a *world-wide* campaign that aims to raise funds through a number of social activities, chief among them the growing of a moustache in the month of November. Although a global event, the `Movember` campaigns are *localized*; each participating country runs its own campaign. In our analysis we focus on the four English-language local campaigns that yield the

[1]Source: `http://us.movember.com/en/about/vision-goals`

Proceedings of the Sixth International Workshop on Health Text Mining and Information Analysis (Louhi), pages 55–63,
Lisbon, Portugal, 17 September 2015.

most donations via Twitter: the United States, the United Kingdom, Canada, and Australia[2]. Globally, Movember can be considered a success, as in 2013 alone (the year we investigate) funds in excess of 123 million *AU$* were raised world-wide[3].

In our work we investigate whether social media activities can *explain* the success of the campaign (both in terms of raising awareness and financially) by correlating Twitter usage with Movember website visits and received donations. We chose Twitter as our social media channel of choice, due to its popularity and ubiquitous nature in the English-speaking world. We investigate the differences and similarities between the Movember Twitter campaigns running in different countries, and aim analyze to what extent those factors can explain awareness and fund-raising metrics.

In the remainder of this paper we first discuss previous findings concerning social media-based health campaigns (§2), before introducing the research hypotheses we focus on in this work and the necessary data sources (§3). Our results are discussed in §4. Lastly, we outline potential avenues for future work in §5.

2 Health Campaigns & Social Media

In this section we provide an overview of existing health campaign research across social media channels. Almost all research conducted in this area investigates the social media portal Twitter. An overview of the employed data in past works is presented in Table 1.

Thackeray et al. (2013) analyzed the impact of the *Breast Cancer Awareness* month (an international campaign held annually in October) on Twitter users. They focused on engagement metrics and found that tweets discussing breast cancer issues spiked dramatically in the beginning of October but quickly tapered off. In terms of topical aspects, organizations and celebrities posted more often than individuals about fundraisers, early detection and diagnoses, while individuals focused more on wearing pink[4]. Similarly, a topic analysis was conducted by Bravo and Hoffman-Goetz (2015) on the 2013 Canadian Movember campaign. The authors categorized 4,222 sampled tweets related to the campaign into four different categories (health information, campaign, participation and opinion). Due to the small number of identified health information tweets in the sample (considered to be the main signal of increased awareness), the authors concluded that the goal of raising awareness has not been met.

Lovejoy et al. (2012) investigated how non-profit organizations *use* Twitter by analyzing more than 70 different organizations, among them 19 health care organizations, along various basic aspects including the number of followers, tweets, retweets, etc. Importantly, the authors found that most organizations use Twitter as a one-way communication channel instead of making full use of its potential and multi-way communication. Smitko (2012) developed two theories, of how non-profit organizations can build and strengthen their relationships with donors on Twitter: the *Social Network Theory* (SNT) and the *Social-Judgement Theory* (SJT). According to SNT, organizations need to strengthen their network of trust by engaging more with their followers while in SJT, organizations need to tailor the content of their tweets to match the interest of their followers. Due to the small-scale nature of the empirical analysis (based on 300 tweets), we consider it an open question to what extent those theories hold.

While to our knowledge, no existing work has considered the financial success of health campaigns, we note that Sylvester et al. (2014) studied the relationship between social media activities (on Twitter and news streams) and donations to a large non-profit organization during hurricane Irene, a tropical cyclone that hit the US in 2011. A spatial analysis revealed that donors living in states affected by Irene donated more than donors in non-affected states.

To summarize, past works have shown that (i) various types of social media users behave differently during health campaigns (celebrities vs. individuals vs. organizations), and (ii) sufficient content related to health campaigns is created on Twitter. What we are lacking is a large-scale analysis of the impact these social media health campaigns have across countries and on fund-raising.

3 Tweets & Donations of Movember

One goal of our work is to establish whether we can *explain* donations the local Movember cam-

[2]Note that these four countries are also in the overall top-five countries in terms of donations.

[3]Source: http://us.movember.com/about/annual-report

[4]A pink ribbon is the symbol of the campaign.

Article	Campaign/Event	Data	Processing	Main Result(s)
(Bravo and Hoffman-Goetz, 2015)	- Movember - Nov. 2013 - Canada	22.3K tweets containing #Movember and located in Canada (user-profile based)	Content analysis	Tweets discussing health topics are significantly outnumbered by tweets discussing non-health topics.
(Sylvester et al., 2014)	- Hurricane Irene - Aug./Sep/ 2011 - United States	- 22K geotagged tweets containing keywords related to Irene - 10K mobile donations - 28K Web donations	Spatial and temporal analysis	- The number of tweets correlate positively with the number of Web donations. - Mobile donations are mostly caused by the relief agency's text message solicitation - Users directly affected by the hurricane display greater social media activity and donate more often
(Thackeray et al., 2013)	- Breast Cancer Awareness - Sep.-Dec. 2012 - N/A	1.3M tweets containing breast cancer related keywords	Content analysis	- Tweets spiked dramatically the first few days of the campaign. - Organizations & celebrities emphasized fund-raisers, early detection, and diagnoses; individuals focused on wearing pink.
(Lovejoy et al., 2012)	- 73 non-profit organizations - Nov.-Dec. 2009 - United States	4.6K tweets posted by organizations	User categorization	Organizations use Twitter mostly as one-way communication channel
(Smitko, 2012)	- 2 health care non-profit & 1 for-profit organizations - 12 hours on Feb. 8, 2011 - Canada	300 tweets either posted by the organizations or mentioning them	Content analysis	Categorized the style of communication into two types: Social Network Theory and Social Judgment Theory

Table 1: Overview of data sets employed in previous work.

paigns received through Twitter[5]. We are thus conducting an exploratory analysis on two distinct data sources:

Twitter Corpus Tw_{Mov}**:** This corpus contains **all** tweets[6] published during the month of November 2013 that contain the keyword *Movember* — 1,113,534 tweets in total, posted by 688,488 unique Twitter users across the world. Twenty-one local `Movember` campaign accounts are active, such as `@MovemberUK`, `@MovemberAUS` and `@MovemberCA`. To enable a country-by-country analysis, we estimated the country each tweet was sent from, according to the machine learning approach described by Van der Veen et al. (2015). In this manner, we were able to label all tweets in our data set with the (likely) country of origin. The approach has been shown to have a country-level accuracy above 80%, a level we consider sufficiently high for our purposes. In total, tweets from 125 different countries were found. The geographic distribution of these tweets is presented in Figure 1, normalized with respect to each country's population, to allow a comparison across countries. It is evident, that the `Movember` campaign is most popular in North America, Australia and Europe. Most activity (relative to the population) is generated by Twitter users in the UK, followed by those in Canada. Thus, the four countries we focus our analysis on are not only among the most active in terms of fund-raising, but also among the most active in terms of `Movember`-related Twitter usage.

`Movember` data: The `Movember` website visitor and donation data we gathered from 2013 is restricted to those visitors and donations the individual national `Movember` campaigns received through Twitter. Overall, in 2013, 357,400 *AU$* were donated through Twitter, spread over 21 national campaigns (though donations were received from 179 countries in total). Thus, only 2.9% of all 2013 donations were received through Twitter. This is a limiting factor to our work, but at the same time allows us to be certain that all of our `Movember` website visitors and donors were exposed to Twitter activities related to `Movember`. Our data set has a single day resolution with all of the following information being available for each individual national campaign website: (1) the

[5]Defined as donations received from users that clicked on a donation link on Twitter.

[6]Twitter provided access to their firehose for this study.

Figure 1: Geo-spatial distribution of all tweets in Tw_{Mov}. We normalized the number of tweets originating in each country by each country's population.

number of visitors, (2) the number of returning visitors, (3)-(4) the number of financial transactions from new and returning visitors, and, (5)-(6) the number of total revenue generated from new and returning visitors. Note that this data does not contain information identifying individual users, it is an aggregate — per day — of all user activities on each `Movember` campaign website. For the four national campaigns investigated in this work, the visitors and donations are listed in Table 3.

As already indicated, `Movember` is a social event, members of the campaign are called *Mo Bros* (men) and *Mo Sistas* (women). Every member can register on the `Movember` website and collect donations through that site (localized per country). *Mo Bros* & *Mo Sistas* can join to form teams and fund-raise together. While growing a moustache is the most common activity, *Mo Bros*/*Mo Sistas* can also use alternative social activities for fund-raising. At the end of the one-month campaign cycle, the teams and individuals raising the most donations within their country receive awards and prices.

3.1 Research Hypotheses

Based on our research goal, we developed three research hypotheses:

H1: The more well-known Twitter users (celebrities and organizations) support a `Movember` campaign, the more awareness and funds the campaign will raise.

H2: `Movember` campaigns that emphasize the social and fun aspect of the campaign, *engage* the users better and thus will raise more awareness and funds.

H3: `Movember` campaigns that focus on health topics, raise more *awareness* to the campaign and thus will raise more funds.

H2 and **H3** are competing hypotheses, as prior works have not offered conclusive evidence to emphasize one direction (health vs. social) over another.

3.2 From Hypotheses to Measurements

Having presented the research hypotheses that guide our work, we now describe how to empirically measure to what extent they hold.

Based on the `Movember` data set, we can directly measure the impact on donations. At the same time though, we cannot directly measure awareness; we chose to approximate this metric by the number of visitors the `Movember` website receives.

To examine **H1** we require a definition for what constitutes a well-known Twitter user (a "celebrity"). We start with the definition posed by Thackeray et al. (2013), according to which celebrities have more than $f_{USA} = 100,000$ followers and are verified by Twitter. As this definition was derived for tweets originating in the United States, we normalize $f_{Country}$ according to the country's population and remove the requirement of being verified. Specifically, for the remaining three countries we employ the following cutoffs: $f_{Canada} = 11,000$, $f_{UK} = 20,000$, and $f_{Australia} = 7,000$.

To investigate the impact of health (related) organizations on Twitter, we define *health organizations* as those Twitter accounts with more than $5,000$ followers and at least one of the following keywords in their Twitter profile (an approach borrowed from (Thackeray et al., 2013)): {*cancer, health, pharmacy, pharmaceutical, campaign, government, firm, company, companies, news, group, society, committee, volunteer, we, official, marketing, promotions and forum*).}. The overlap between both types of users (well-known vs. organizations) is between 2.2% (US) and 30.7% (Australia).

3.2.1 Manual Annotation Efforts

Hypotheses **H2** and **H3** require a content analysis of the Twitter messages. For this purpose, one of the authors manually annotated 2,000 randomly drawn English-language tweets (with 500 tweets each drawn from the UK, Canada, the United States and Australia) from Tw_{Mov} into several categories, inspired by the work of Bravo and Hoffman-Goetz (2015). We distinguish five main categories: *health*, *campaign*, *participation*, *social* and *other*, with each one (except *other*) containing between two and three sub-categories (e.g. *health* tweets are further categorized as *cancer*, *general* and *mental*). Overall, we distinguish 12 different categories/sub-categories. Tweets can belong to multiple categories or sub-categories; tweets that are not found to belong to any of the first four categories are classified as *other*. An overview of the categories and the resulting annotations (including examples of categorized tweets) is shown in Table 2. Across all countries, we find the social aspect to be the most pronounced in our sample — 51% of the sampled tweets are categorized as such. Less than 5% of the tweets mention health issues and even more strikingly, the second pillar of `Movember`'s campaign (mental health) is almost completely absent in our sample. These results are largely in line with Bravo and Hoffman-Goetz (2015)'s findings for the Canadian `Movember` campaign, where cancer-related tweets were found in only 0.6% of the sample. This manual annotation effort does not only serve as a confirmation of (Bravo and Hoffman-Goetz, 2015), it also shows that these findings hold *across* countries.

3.2.2 Automatic Classification

Due to the small number of manually annotated tweets in the individual sub-categories, we decided to automatically classify all tweets of Tw_{Mov} according to the most opposing ends of the spectrum: *health* vs. *social*. This was done separately for each country. Concretely, we aim to classify each tweet into one of four categories: (1) *health*, (2) *social*, (3) *health & social* or (4) *other*. In order to add robustness to the classifier, we use the insights gained during the manual annotation process to enlarge our training set by automatically selecting additional positive training tweets. For the *health* classifier, tweets containing one of the following key phrases were used: {*prostate, testicular, cancer, mental, health*}. Similarly, for the *social* classifier, we relied on tweets containing at least one of: {*gala, party, event, contest, competition, stach, handlebar, facial hair, shave, instagram, twitter.*photo.*} as positive training data. Recall, that all tweets in our corpus contain the term *Movember* by definition, thus ensuring topicality. Overall, in this manner we labelled 406,709 tweets across all countries, consisting of 120,601 *health* tweets and 286,108 *social* tweets. A total of 35,489 tweets were identified as being both *social* and *health*-related. These simple rules have thus allowed us to categorize 36.5% of all tweets in Tw_{Mov}; the remaining 63.5% of tweets are categorized according to our classifier output.

We train separate classifiers for each country. We randomly draw 5,000 labelled health (social) tweets as positive training examples of the health (social) classifier. We draw the same amount of non-health (non-social) tweets as negative training examples for balanced training[8]. We performed basic data cleaning steps, removing stopwords (which in this case includes the term "Movember") and employing stemming. As classification algorithm we selected Naïve Bayes with terms as features[9]. We classified the tweets in Tw_{Mov} to zero, one or both categories (health/social) depending on the confidence threshold of the individual classifier (a tweet classified with confidence ≥ 0.5 is assigned to the classifier's category).

4 Results

To determine the influence on the number of donations and visitors, we correlate (using Pearson's correlation coefficient r) the Twitter-based metrics (e.g. number of tweets) with the donation and visitor data from the `Movember` data set on a day-by-day basis for the month of November.

[8]Note that using all already labelled tweets as positive training examples is not possible, as in effect nearly all remaining tweets would act as negative training examples in a balanced training setup.

[9]We employed the WEKA toolkit: `http://www.cs.waikato.ac.nz/ml/weka/`

	Canada	United States	United Kingdom	Australia
health:cancer	21	16	18	20
East Devon District Council working to raise awareness of male cancers and support cancer research! uk.movember.com/team/1242769 via @movemberuk				
health:general	41	88	45	40
This month, BeTheBoss.ca will be participating in Movember to raise money for medical research, help those living... `fb.me/M4FmLKff`				
health:mental	3	2	0	1
Trott you are a legend. Help support all men going through mental health sturggles. Support my mo! #Movember mobro.co/mrdixon				
%health overall	13.0%	21.2%	12.6%	12.2%
campaign:value	41	69	71	36
I've enlisted in #Movember to change the face of men's health. Donate & join the good fight mobro.co/Perthpotter				
campaign:news	25	18	13	14
Indian man unsure what the Movember fuss is all about panarabiaenquirer.com/wordpress/indi				
campaign:status	28	46	58	24
10 'Mo' days of Movember to go				
%campaign overall	18.8%	26.6%	28.4%	14.8%
participation:support	127	155	96	82
My Wonderful Husband is growing a #Mo for #Movember! Please donate big so I'm living with a hairy man for a reason! mobro.co/mrcaseytalbot				
participation:report	20	32	27	14
Thank You. So far, $535 has been raised for my Mo in Movember. Great result but there is still time to donate - mobro.co/tonylapila				
%participation overall	29.4%	37.4%	24.6%	19.2%
social:moustaches	182	202	202	217
RT @itsWillyFerrell: With great mustache, comes great responsibility. #NoShaveNovember #Movember				
social:service/goods	31	13	30	30
Making mustache chocolate cookies in preparation for my #movember kickoff at work on Monday. #yum!				
social:events	38	35	19	14
RT @SurreyTavern1: Come to our End of #Movember #MoParty - `facebook.com/events/1681280` - great fun and for a good cause! #LiveMusic #Norwich				
%social overall	50.4%	50.0%	50.2%	52.2%
other	91	77	118	76
happy #Movember				

Table 2: Overview of the manual annotation results. For each country, 500 tweets are sampled and categorized. For each sub-category, an example tweet from our corpus is shown.

	Donation (AU$[7])	Transactions	Users/Visitors	Population
Canada	91,741	2,054	43,720	35 M
United States	79,828	1,847	76,257	321 M
United Kingdom	75,124	4,397	95,867	65 M
Australia	13,170	583	11,194	24 M
in total	284,897	8,955	229,745	—

Table 3: Overview of the 2013 `Movember` campaign donations received through Twitter. The final column lists each country's population (in millions).

Hypothesis H1: To investigate **H1**, we correlate the number of `Movember` tweets by well-known Twitter users on a given day with the donations/visitors to the `Movember` campaign website on a per-country basis. The results are shown in Table 4. While the visitors correlate to a significant degree with several tweet-based measures for the United Kingdom and Australia, we do not observe significant correlations for visitors in Canada or the US. Organizations have a similar impact to Twitter celebrities (normalized by country) in terms of drawing visitors to the `Movember` website. Contrary to our intuition, we do not observe any significant correlations between the daily number of donations and Twitter activities.

	Canada	United States	United Kingdom	Australia
Total number (#) of tweets	81,614	298,720	565,503	24,558
#Tweets by well-known Twitter users	179	1,445	662	39
$r_{donations}$	-0.02	0.13	0.35	0.30
$r_{visitors}$	0.13	0.23	0.36	0.37†
#$f_{Country}$ normalized tweets	2,056	1,445	6,167	2,158
$r_{donations}$	-0.05	0.13	0.19	0.56‡
$r_{visitors}$	0.22	0.23	0.58‡	0.68‡
#Organizational tweets	17,535	50,131	78,174	5,222
$r_{donations}$	-0.04	0.10	0.14	0.47‡
$r_{visitors}$	0.27	0.33	0.56‡	0.77‡

Table 4: Overview of the number of Tw_{Mov} tweets across the month of November 2013 as well as their correlation (day-by-day) with the number of daily donations and daily visitors to each country's Movember website. The thresholds for statistical significance (for $N = 30$ days) are † $r = 0.37$ ($p < 0.05$) and ‡ $r = 0.47$ ($p < 0.01$) respectively.

Hypotheses H2 & H3 In Table 5 we present the impact social and health topics have on Movember donations and visitors. The results are similar to the previous experiment: we observe significant correlations only with Movember visitor data; Australia & United Kingdom exhibit moderate to strong correlations while for Canada & the US the correlations are weak to non-significant. Considering the influence of health vs. social we find that social tweets exhibit a stronger correlation with visitor data than health tweets across all countries — this in fact is the only experiment where statistically significant results are observed across all four countries.

Further Insights In Figure 2 we visualize the relationship between the number of visitors/donations and the number of health/social tweets in the form of scatter plots. While the visitor data shows few outliers (corresponding to the first & last day of the campaign) and has a clear linear trend, the donation plot is evidently non-linear without a clear pattern emerging.

Finally, in Figure 3, we plot — exemplary for the United Kingdom — the overall trends in the number of tweets, the number of Movember visitors and the number of Movember donations between the end of October 2013 and early December 2013. We observe that over time, the overall tweet volume declines slightly (apart from the final day of the campaign), while the number of visitors and the number of donations are in a reverse relationship: the number of visitors steadily declines over the month of the campaign while the number of donations steadily increases. Twitter activity related to Movember quickly ceases to exist after the end of November.

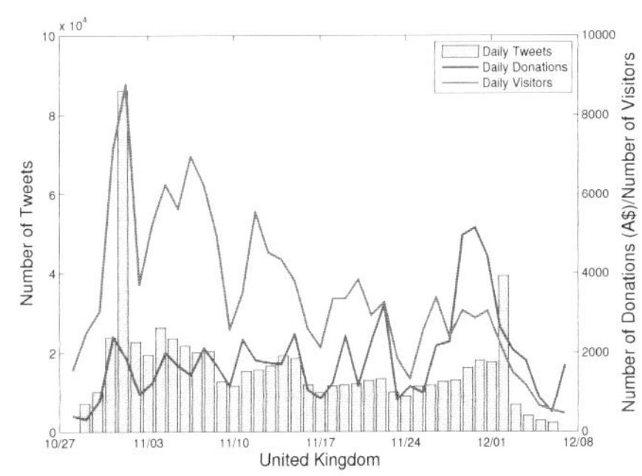

Figure 3: Daily trends in the United Kingdom: overview of the number of tweets, visitors, and number of donations. The timeline starts on October 27, 2013 (10/27) and ends on December 8, 2013 (12/08).

5 Conclusions

In this paper, we investigated the impact of different social media strategies on a health campaign's ability to raise awareness and attract funds. We investigated the specific use case of Movember, a global campaign which enjoys widespread popularity in many countries. We focused our analyses on the four most active English-language countries of the Movember campaign.

Our findings partially corroborate previous findings on raising awareness, especially those in (Bravo and Hoffman-Goetz, 2015), while expanding on them across several dimensions, most importantly the number of countries investigated and the size of the investigated social media

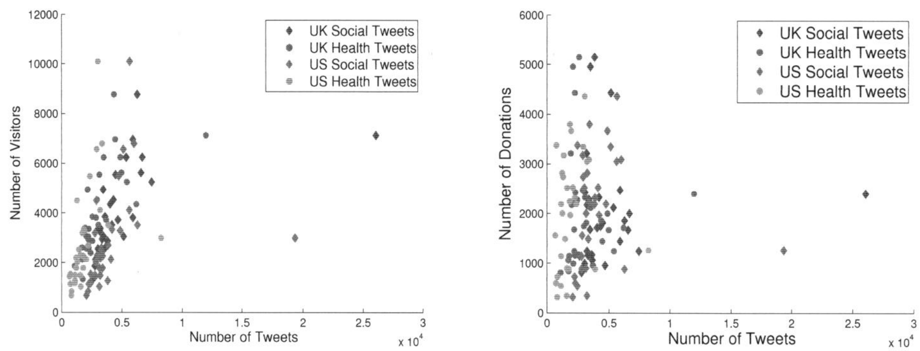

(a) Visitors

(b) Donations

Figure 2: Scatter plots of the daily number of health / social tweets and the daily number of visitors / donations shown exemplary for the United Kingdom and the United States.

	Canada	United States	United Kingdom	Australia
#English tweets	78,382	287,479	515,605	24,189
$r_{donations}$	-0.09	0.05	0.09	0.32
$r_{visitors}$	0.24	0.30	0.56‡	0.80‡
#Classified as *health* tweets	13,360	58,283	96,000	5,014
$r_{donations}$	-0.09	0.06	0.07	0.30
$r_{visitors}$	0.21	0.30	0.55‡	0.75‡
#Classified as *social* tweets	28,594	124,954	149,226	13,010
$r_{donations}$	-0.13	0.11	-0.02	0.08
$r_{visitors}$	0.38†	0.43†	0.68‡	0.83‡

Table 5: Overview of the number of tweets classified according to their health and/or social intent as well as their correlation (day-by-day) with `Movember` donation and visitor data. The thresholds for statistical significance (for $N = 30$ days) are † $r = 0.37$ ($p < 0.05$) and ‡ $r = 0.47$ ($p < 0.01$) respectively.

sample. We find that across countries Twitter users mostly focus on the social aspect of the `Movember` campaign, with relatively few tweets focusing on the health aspect of `Movember`. Additionally, those users that do mention health-related issues, often use generic statements, instead of focusing on the two specific health issues that `Movember` aims to address (cancer and mental health). Surprisingly, the mental health aspect of `Movember` is virtually not discussed at all.

To explore the impact of social media strategies on awareness and fund-raising, we analysed the relationship between `Movember` website visitor & donation data and Twitter activities. We found significant correlations between `Movember` visitors and the `Movember`-related activities of well-known Twitter users. We also found clear evidence that social tweets have a higher impact on visitors than health tweets. While the observed correlations were moderate to strong for the United Kingdom and Australia, we only found weak to non-significant correlations for Canada and the United States. Across all countries, we did not find significant correlations between donations and Twitter activities.

Based on these findings, we plan to investigate on a more fine-grained and semantic level in what aspects the Twitter-based `Movember` activities differ between Australia/UK and Canada/US. We will also consider a temporal analysis of the donation/visitor data, comparing trends across several years of `Movember` donation data and Twitter activities. We also intend to incorporate more fine-grained information about the Twitter users in our analyses, such as their motivations to participate in the campaign (Nguyen et al., 2015).

Acknowledgments

This research was funded in part by the 3TU Federation and the Dutch national projects COMMIT and FACT. We are grateful to Twitter and Movember for providing the data.

References

Harshavardhan Achrekar, Avinash Gandhe, Ross Lazarus, Ssu-Hsin Yu, and Benyuan Liu. 2011. Predicting flu trends using twitter data. In *Computer Communications Workshops (INFOCOM WKSHPS), 2011 IEEE Conference on*, pages 702–707. IEEE.

Justin C Bosley, Nina W Zhao, Shawndra Hill, Frances S Shofer, David A Asch, Lance B Becker, and Raina M Merchant. 2013. Decoding twitter: Surveillance and trends for cardiac arrest and resuscitation communication. *Resuscitation*, 84(2):206–212.

Caroline A Bravo and Laurie Hoffman-Goetz. 2015. Tweeting about prostate and testicular cancers: Do twitter conversations and the 2013 movember canada campaign objectives align? *Journal of Cancer Education*, pages 1–8.

Cynthia Chew and Gunther Eysenbach. 2010. Pandemics in the age of twitter: content analysis of tweets during the 2009 h1n1 outbreak. *PloS one*, 5(11):e14118.

Aron Culotta. 2010. Towards detecting influenza epidemics by analyzing twitter messages. In *Proceedings of the first workshop on social media analytics*, pages 115–122. ACM.

Ernesto Diaz-Aviles and Avaré Stewart. 2012. Tracking twitter for epidemic intelligence: case study: Ehec/hus outbreak in germany, 2011. In *Proceedings of the 4th Annual ACM Web Science Conference*, pages 82–85. ACM.

Kristen Lovejoy, Richard D Waters, and Gregory D Saxton. 2012. Engaging stakeholders through twitter: How nonprofit organizations are getting more out of 140 characters or less. *Public Relations Review*, 38(2):313–318.

S Anne Moorhead, Diane E Hazlett, Laura Harrison, Jennifer K Carroll, Anthea Irwin, and Ciska Hoving. 2013. A new dimension of health care: systematic review of the uses, benefits, and limitations of social media for health communication. *Journal of medical Internet research*, 15(4).

Dong Nguyen, Tijs A. van den Broek, Claudia Hauff, Djoerd Hiemstra, and Michel Ehrenhard. 2015. #SupportTheCause: Identifying motivations to participate in online health campaigns. In *Proceedings of EMNLP 2015*.

Daniel Scanfeld, Vanessa Scanfeld, and Elaine L Larson. 2010. Dissemination of health information through social networks: Twitter and antibiotics. *American journal of infection control*, 38(3):182–188.

Isabel Segura-Bedmar, Ricardo Revert, and Paloma Martínez. 2014. Detecting drugs and adverse events from spanish health social media streams. In *Proceedings of the 5th International Workshop on Health Text Mining and Information Analysis (Louhi)@ EACL*, pages 106–115.

Kate Smitko. 2012. Donor engagement through twitter. *Public Relations Review*, 38(4):633–635.

Jared Sylvester, John Healey, Chen Wang, and William M Rand. 2014. Space, time, and hurricanes: Investigating the spatiotemporal relationship among social media use, donations, and disasters. *Robert H. Smith School Research Paper No. RHS*, 2441314.

Rosemary Thackeray, Scott H Burton, Christophe Giraud-Carrier, Stephen Rollins, and Catherine R Draper. 2013. Using twitter for breast cancer prevention: an analysis of breast cancer awareness month. *BMC cancer*, 13(1):508.

Han Van der Veen, Djoerd Hiemstra, Tijs van den Broek, Michel Ehrenhard, and Ariana Need. 2015. Determine the User Country of a Tweet. Technical report, TR-CTIT-15-05, Centre for Telematics and Information Technology, University of Twente.

Karl Vance, William Howe, and Robert P Dellavalle. 2009. Social internet sites as a source of public health information. *Dermatologic clinics*, 27(2):133–136.

Exploring Word Embedding for Drug Name Recognition

Isabel Segura-Bedmar, Víctor Suárez-Paniagua, Paloma Martínez
Computer Science Department
University Carlos III of Madrid, Spain
{isegura,vspaniag,pmf}@inf.uc3m.es

Abstract

This paper describes a machine learning-based approach that uses word embedding features to recognize drug names from biomedical texts. As a starting point, we developed a baseline system based on Conditional Random Field (CRF) trained with standard features used in current Named Entity Recognition (NER) systems. Then, the system was extended to incorporate new features, such as word vectors and word clusters generated by the Word2Vec tool and a lexicon feature from the DINTO ontology. We trained the Word2vec tool over two different corpus: Wikipedia and MedLine. Our main goal is to study the effectiveness of using word embeddings as features to improve performance on our baseline system, as well as to analyze whether the DINTO ontology could be a valuable complementary data source integrated in a machine learning NER system. To evaluate our approach and compare it with previous work, we conducted a series of experiments on the dataset of SemEval-2013 Task 9.1 Drug Name Recognition.

1 Introduction

The automatic recognition of biomedical entities from scientific texts can markedly reduce the time that experts spend populating biomedical knowledge bases and annotating papers and patents. Furthermore, Named Entity Recognition (NER) is a crucial component for many Natural Language Processing (NLP) systems such as relation extraction, text classification or sentiment analysis systems, among many others.

Conditional Random Fields (CRF) often show best results in the recognition of drugs and chemical names (Krallinger et al., 2015a; Segura Bedmar et al., 2013). So far the most popular features for CRF-based NER systems concern syntactic and semantic properties of words (such as tokens, part-of-speech (POS) tags, lemmas, orthographic and lexicon features, among others). In this work, we develop a system based on a CRF to recognize drug mentions occurring in the DDI corpus (Herrero-Zazo et al., 2013)[1]. It consists of two different datasets: DDI-DrugBank (792 texts selected from the DrugBank database) and DDI-MedLine (233 MedLine abstracts on the subject of DDIs). This corpus will allow us to compare our system to the participating systems in the SemEval-2013 Task 9.1 DrugNER Task.

One of the goals of this paper is to study whether the DINTO ontology[2] (Herrero Zazo, 2015) can provide valuable information for this task. As far as we know, DINTO is the first ontology providing a comprehensive and accurate representation of drug-drug interactions (DDI) knowledge. The DINTO ontology contains a total of 25,809 classes, in particular 8,786 drugs and 11,555 DDIs. Several domain resources such as the CheBI ontology (Degtyarenko et al., 2008), the DrugBank database (Wishart et al., 2006) or the OAE ontology (He et al., 2014) have been reused to create DINTO. Furthermore, it was designed to be used by the computer science community working on the DDI domain. A detailed description of the DINTO ontology can be found in Herrero-Zazo's PhD thesis (Herrero Zazo, 2015).

As the main contribution, this work explores the effectiveness of new features for the Drug NER task, in particular, word clusters and word vectors generated using the Word2Vec tool (Mikolov et al., 2013a), a word embedding model based on a neural network (NN). We hypothesize that the use

[1] http://labda.inf.uc3m.es/ddicorpus
[2] http://www.obofoundry.org/cgi-bin/detail.cgi?id=DINTO

Proceedings of the Sixth International Workshop on Health Text Mining and Information Analysis (Louhi), pages 64–72,
Lisbon, Portugal, 17 September 2015.

of word embedding features would allow us to accurately detect even those drugs that are not in the training set or in the DINTO ontology. A word embedding is a function to map words to high-dimensional vectors. At present, NN is one of the most used learning techniques for generating word embeddings (Mikolov et al., 2013b). The essential assumption of word embedding is that semantically close words will have similar vectors. Word embeddings have shown promising results in NLP tasks, such as named entity recognition, sentiment analysis or parsing (Turian et al., 2010; Socher et al., 2013a; Socher et al., 2013b). However, to the best of our knowledge, this technique has hardly ever been exploited in drug name recognition (Liu et al., 2015).

In fact, our work is the first to explore the word embedding potential using the whole word2vec vector for drug name recognition. In contrast to (Liu et al., 2015), we also train the word embedding features (word clusters and word vectors) using the latest wikipedia dump[3], which contains more than 3 billion words, as well as the 2013 release of MedLine[4], which they used for genereting their word representations. This release contains approximately one million words, being thus much smaller than the Wikipedia collection. While MedLine is a biomedical literature database, Wikipedia covers many different domains of knowledge. However, we believe that the larger the dataset used for training the Word2Vec models, the better word embeddings should be obtained. Thus, we would like to compare the effectiveness of word embeddings features trained on a specific domain corpus, such as MedLine, to those trained on a larger collection, such as Wikipedia.

Another key difference of our work with (Liu et al., 2015) is that while they only gave results for the whole DDI corpus, we analyze and discuss the effect of the DINTO and word2vec features on each one of the datasets: DDI-DrugBank and DDI-MedLine. This analysis is necessary in order to know what features are more efficient on each dataset. MedLine abstracts are very different from DrugBank texts. While abstracts are mainly addressed to scientists in life sciences, texts from DDI-DrugBank are written in a language understandable to patients.

The paper is organized as follows. In the next section, we introduce the two main shared tasks for drug name recognition task organized so far: the BioCreative IV ChemdNER task and the drugNER subtask of the SemEval-2013 DDIExtraction challenge. Section 3 describes the datasets used and the experiments performed. The experimental results are presented and discussed in Section 4. We conclude in Section 5 with a summary of our findings and some directions for future work.

[3]http://dumps.wikimedia.org/

[4]http://www.nlm.nih.gov/databases/journal.html

2 State of the art

2.1 CHEMDNER task

The BioCreative IV CHEMDNER (Chemical compound and drug name recognition) task was devoted to NER focusing on detecting chemical entity mentions. Twenty-six teams participated in this task and as a result a corpus containing 10,000 PubMed abstracts annotated with 84,355 chemistry and chemical entity mentions was generated (Krallinger et al., 2015b). An overview of the task as well as of the main relevant characteristics of participating systems is given in (Krallinger et al., 2015a).

Participating systems used three approaches to recognize chemical entity mentions: (a) supervised machine learning techniques (used by 17 systems). CRF was the most used technique followed by Support Vector Machines (SVM) and logistic regression. These systems used different types of features: word level features (such as ngrams, numerical items and digits, word length, part-of-speech, among others), lookup features extracted from dictionaries and gazetteers and document features (for example, coocurrences of mentions); (b) rule-based approaches are used in two systems in the form of lexical patterns that implements the IUPAC nomenclature guidelines to detect formulas or specific sequences of compounds (this strategy requires a high understanding of chemical naming standards as well as annotation guidelines) and (c) dictionary-based approaches are integrated in four systems where domain-specific resources (such as CheBI[5], PubChem[6] or DrugBank[7]) and gazetteers are expanded with lexical variations to improve recall scores taking into account that a post-processing task of removing and pruning lexical entries is required. Only three

[5]https://www.ebi.ac.uk/chebi/

[6]https://pubchem.ncbi.nlm.nih.gov

[7]http://www.drugbank.ca

systems tried a hybrid approach combining machine learning and rule-based strategies. Analyzing the runs submitted by participating teams, it is important to highlight that the top ranked system (Leaman et al., 2015) (87,39% of F-score) implemented a hybrid approach that combines a CRF model, a set of patterns to identify special types of mentions and gazetteers. This score is very close to the inter human annotator agreement (IAA) in this task (91%).

2.2 SemEval-2013 DrugNER task

The DDIExtraction Shared Task 2013 (Segura Bedmar et al., 2013; Segura-Bedmar et al., 2014) is the second edition of the DDIExtraction Shared Task series, a community-wide effort to promote the implementation and comparative assessment of NLP techniques in the field of the Pharmacovigilance domain. To attain this aim, two main tasks were proposed: the recognition of pharmacological substances (DrugNER task) and the detection and classification of drug-drug interactions (DDI task) from biomedical texts. Four types of pharmacological substances were defined: *drug* (generic drug names), *brand* (branded drug names), *group* (drug group names) and *drug-n* (active substances not approved for human use). The results of the participating systems were evaluated according to four evaluation criteria: strict (which demands exact boundary and entity type matching), exact (which only demans exact boundary matching), partial (which only demands partial boundary matching) and type (which demands partial boundary and entity type matching).

A total of 6 teams participated in the DrugNER subtask. The reader can find the full ranking information in (Segura Bedmar et al., 2013). In general, the results on the DDI-DrugBank dataset were much better than those obtained on the DDI-MedLine dataset. While DDI-DrugBank texts focus on the description of drugs and their interactions, the main topic of DDI-MedLine texts would not necessarily be on DDIs. Coupled with this, it is not always trivial to distinguish between substances that should be classified as pharmacological substances and those that should not. This is due to the ambiguity of some pharmacological terms. For example, *insulin* is a hormone produced by the pancreas, but can also be synthesized in the laboratory and used as drug to treat insulin-dependent diabetes mellitus. The participating systems should be able to determine if the text is describing a substance originated within the organism or, on the contrary, it describes a process in which the substance is used for a specific purpose and thus should be identified as pharmacological substance.

The best results were achieved by the WBI team (Rocktäschel et al., 2013) with a CRF algorithm. The system employed a domain-independent feature set along with features generated from the output of ChemSpot (Rocktäschel et al., 2012), an existing chemical named entity recognition tool, as well as a collection of domain-specific resources. Its model was trained on the training dataset as well as on entities of the test dataset for the DDI task. In the detection subtask (which only requires exact boundary matching), this system achieved an F1 of 90% on the DDI-DrugBank dataset and an F1 of 78% on DDI-MedLine. As expected, the results of the classification subtask (strict evaluation) were worse, showing an F1 of 87.8% on DDI-DrugBank and 58.1% on DDI-MedLine.

3 Method

This section describes the datasets and settings used in our experiments.

3.1 Datasets

The major contribution of DDIExtraction was to provide a benchmark corpus, the DDI corpus. The corpus was manually annotated with a total of 18,502 pharmacological substances and 5,028 DDIs. It consists of two different datasets: DDI-DrugBank (792 texts selected from the DrugBank database) and DDI-MedLine (233 MedLine abstracts on the subject of DDIs). A detailed description of the DDI corpus can be found in (Herrero-Zazo et al., 2013).

The corpus was split in order to build the datasets for the training and evaluation of the different participating systems. Approximately 77% of the DDI corpus documents were randomly selected for the training dataset and the remaining was used for the test dataset. The training dataset is the same for both subtasks since it contains entity and DDI annotations. The test dataset for the DrugNER task was formed by discarding documents which contained DDI annotations. Entity annotations were removed from this dataset to be used by participants. The remaining documents (that is, those containing some interactions) were used to create the test dataset for the DDI task. Since entity annotations are not removed from these documents, the test dataset for the DDI task can also be used as additional training data for the DrugNER task.

Table1 shows the basic statistics on the training and test datasets for the DrugNER task.

3.2 Experiments

As it stated in the previous section, most successful approaches for drug name recognition have used machine learning algorithms such as CRFs trained with linguistic features (tokens, lemmas or POS tags, among others) and semantic features from domain resources such as ontologies or dictionaries. Encouraged by the good results of the CRF-based methods, we propose a system based on CRF and also explore word embedding features provided by the Word2vec tool. In particular, we used a python binding[8] to CRFsuite (Okazaki, 2007).

CRF performs the NER task as a classification task on each token, determining whether it is an entity or not. To represent the class of each token, we used the BIO tagging scheme. According to this scheme, each token is tagged as either beginning entity token (B), inside entity token (I) or outside token (O). For the detection subtask (exact criterion), we only considered three classes: B-ENTITY, I-ENTITY and O. However, since we had to classify four different types (drug, brand, group and drug-n), we used nine different classes for the classification task.

As a first stage, we developed a baseline system using a CRF algorithm in which each token is represented with the following features:

- The context window of three tokens to its right and to its left in the sentence. The context window also includes the current token.
- POS tags and lemmas in the context window are also considered.
- An orthography feature which can take the following values: upperInitial (the token begins with an uppercase letter and the rest are lowercase), allCaps (all its letters are uppercase), lowerCase (all its letters are lowercase) and mixedCaps (the token contains any mixture of upper and lowercase letters).
- A feature representing the type of token: word, number, symbol or punctuation.

As one of our goals is to study the contribution of DINTO in the task, in a second stage, we also considered a binary feature that indicated whether the current token was found in the DINTO ontology.

Figure 1 shows a pipeline of GATE components used to process the texts and to obtain the feature set used to train the CRF model. There are five main processing modules: sentence splitter, tokenizer, POS tagger, morphological analyzer and the Gate onto root gazetteer, which links text to the DINTO ontology. The ontology is processed to produce a flexible gazetteer taking into account alternative morphological forms of the instances of the ontology.

The main hypothesis of this work is that the incorporating of word embeddings as features into a CRF model could help to recognize unseen or very rare drug mentions in the training set. Thus, we train word embeddings using the Word2vec tool. Word2vec only requires a large corpus of sentences as input dataset in order to generate word vectors by training a NN language model. The NN model is able to learn from the different contexts in which a word appears and then to compute its representation as a vector. In this study, Word2Vec tool was trained on two different corpora. As first option, we used the latest wikipedia dump[9], which contains more than 3 billion words. Then, we used the Word2Vec model trained on Wikipedia to obtain the word vectors for all tokens in the DDI corpus.

Based on distributional hypothesis (Harris, 1954), similar words will have similar vectors because they occur in similar contexts. The word vector for the current token was considered as a new feature into an our CRF system. We tried with different dimensions of vectors (50, 100 and 200) (see Table 3). It should be noted that these word representations could be very valuable input, not only for named entity recognition, but also in many other NLP tasks (POS tagging, word name disambiguation, lexical simplification, etc).

Another important advantage of the Word2vec tool is that contains a utility to compute word clusters using a k-means clustering algorithm. Thus, we also used word cluster as a new feature to represent the current token in our CRF-based system.

[8]http://python-crfsuite.readthedocs.org/en/latest/

[9]http://dumps.wikimedia.org/

		Training + Test for DDI task	Test for DrugNER task
DDI-DrugBank	documents	730	54
	sentences	6648	145
	drug	9715	180
	group	3832	65
	brand	1770	53
	drug_n	124	5
DDI-MedLine	documents	175	58
	sentences	1627	520
	drug	1574	171
	group	234	90
	brand	36	6
	drug_n	520	115

Table 1: Statistics on the training and test dataset for the DrugNER task.

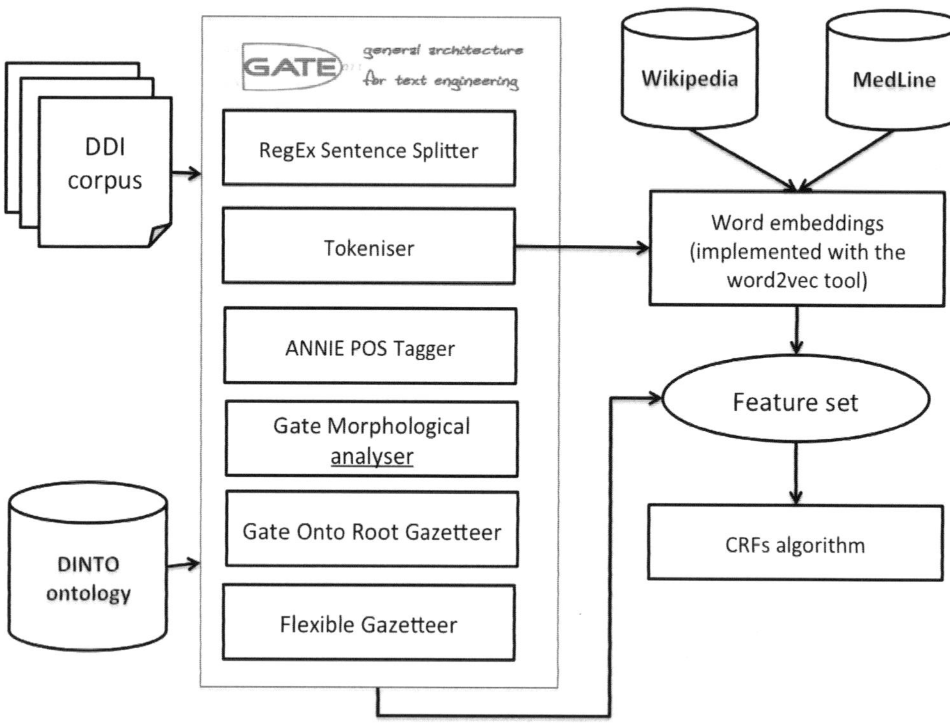

Figure 1: System architecture and pipelines for CRF machine learning-based Drug NER.

Word clusters represent words at a higher level abstraction that may help to recognize even those drug mentions that are not observed in the training set. We performed experiments for different values of k in the k-means (50, 150 and 500). All experiments are summarized in Table 2.

4 Evaluation

Table 3 shows the results for the different settings studied for the detection subtask (exact criterion) and for the classification subtask (strict criterion). The scores correspond to the micro-average values, which were calculated with regarding all classes (B- and I-) of each corresponding subtask.

The following subsections present and discuss the results for each dataset: DDI-DrugBank and DDI-MedLine.

4.1 Results on DDI-DrugBank

4.1.1 Detection subtask

The use of a lexicon feature from DINTO achieved an increase in both precision and recall (and consequently, an improvement of 1% in F1 score).

The results suggest that Word2vec features can potentially lead to improved detection performance. In general, the use of word clusters showed a significant increase in recall values (from 84% to 89%), and hence a gain of 3% in F1. However, word clusters did not seem significantly to alter overall precision values. As expected, word cluster is an effective feature to improve the coverage of the system.

Our initial hypothesis was that Word2vec features trained on MedLine should provide better results because these texts are focused on the biomedical domain, however the results demonstrated that word clusters from Wikipedia, in general, had a better performance than those from MedLine. This may be due to the size of the Wikipedia corpus is significantly larger than the release of Medline used in this work. Therefore, Wikipedia is the best option to train our Word2Vec models in our current settings, though Wikipedia cover a vast array of subjects, not necessarily related to the biomedical domain.

Word cluster features trained on MedLine always seem to provide the same scores, that is, there is no difference between to use a cluster which was calculated using k=50, k=150 or k=500. Word clusters trained on Wikipedia produced better results when the number of clusters is larger. More experiments are necessary to confirm or deny these results. In general, word clusters performed better than word vectors.

To sum up, the results suggest that word clusters are the most influential features for the detection subtask, achieving an improvement of 4% in recall over the baseline system.

4.1.2 Classification subtask

Regarding the results of the classification task on the DDI-DrugBank dataset, the use of Word2vec features did not necessarily give better results than the baseline system and might even be worse (see Table 3). The best F1 (75%) was obtained by five different strategies (see Table 3): baseline, word clusters (k=50) on Wikipedia, word clusters (k=50, k=500) on MedLine and word vectors (d=50) on MedLine.

Similarly, DINTO did not overcome the baseline system yet. Therefore, while the experiments on the detection task show that the use of DINTO and Word2vec features could help to improve the performance, this positive effect does not seem to be present for the classification task.

4.2 Results on DDI-MedLine

4.2.1 Detection subtask

The use of DINTO led to an increase in precision, achieving 10% over the baseline system, and an increase of 3% in recall. Thus, F1-score went up from 61% to 66%.

Word cluster features generated from Wikipedia provided a significant improvement of 6% in recall, but with worse precision than the combination of baseline with DINTO. As was the case on DDI-DrugBank, lower improvements were obtained by the word clusters trained on MedLine. Moreover, word clusters seemed to perform better than word vectors. On the other hand, word vectors trained on MedLine showed precision values very close to those obtained by the baseline system with DINTO.

4.2.2 Classification subtask

Contrary to the evaluation on the DDI-DrugBank dataset, the use of DINTO increased the baseline precision by 8% and the baseline recall by 3%. Therefore, DINTO provide valuable information for the classification of drug entities in scientific texts. This may be due to DINTO incorporates information from several resources such as the ChEBI ontology, the DrugBank database and the ATC classification system[10] (a drug classification system developed by WHO). Word clusters (k=500) achieved the best performance by increasing the recall (by 7%) and thus the F1 accordingly. However, word vectors do not seem to provide an improvement over the results achieved by DINTO.

Although our system does not provide better performance than the WBI system, the use of the DINTO feature show a significant improvement by 9% in precision over the WBI system, but with a sharp reduction in recall.

5 Conclusion

The main contribution of this paper is the incorporation of word embedding features into a CRF-based NER system for drug entities. In addition, we explore if the DINTO ontology can be a valuable resource for the task.

The results suggest that DINTO can lead to improve the performance over the detection subtask. Therefore, we can confirm that the DINTO ontology is a useful resource for the drug name recognition task from scientific texts. For this reason, we intend to continue studying on how to better use DINTO in order to increase the performance of the task. Moreover, we believe that the inclusion of additional semantic features from biomedical resources (such as DrugBank, CheBI, ChemIDPlus, the ATC classification system, Drugs@FD [11], etc) are essential in order to improve performance for the classification subtask.

As we foresaw in the initial hypothesis, Word2vec features achieve a marked improvement in recall for the detection task. Word cluster features trained on Wikipedia seem to provide the most satisfactory results. More experiments are necessary to determine the optimum number of clusters for the task. Although in general our results are not better than those achieved by the top system in the DrugNER task, we strongly believe the use of word embeddings for this task is worth further research.

Our experiments conducted on the DDI corpus allow us to compare our approach with the participating systems of the DrugNER task in the SemEval-2013 DDIExtraction challenge. In general, our system does not perform better than the top system (WBI) in this shared task. However, the results for the classification task on the DDI-MedLine dataset show that DINTO could be a valuable resource to improve precision.

The WBI system provided an F1 of 87.8% on DDI-DrugBank (which is very close to the IAA (0.91)), but performed worse on the DDI-MedLine dataset (showing an F1 of 58.1%). It stands to reason that this system could have already reached the maximum threshold results for the DDI-DrugBank dataset. On the other hand, there is much room for improvement on the DDI-MedLine dataset. The results reported in (Liu et al., 2015) are better than those provided by the WBI system. However, since the authors only provide results for the whole DDI corpus, we cannot know the performance of their system on each dataset and whether their system is able to overcome the WBI system on the DDI-MedLine dataset.

In future work, we will first train the Word2vec tool using a large set of MedLine abstracts. It could provide better results than those obtained from the Word2vec model trained on Wikipedia. Since MedLine is a biomedical literature database, Medline abstracts should provide better word representations for drug entities than those obtained from Wikipedia articles. We also plan to extend the experimentation to the ChemdNER corpus in order to compare our approach to the participating systems of the BioCreative IV CHEMDNER task. We also intend to carry out an error analysis to determine the main causes for wrong detection and classification.

Furthermore, we will still explore additional word embedding features for the drugNER task. In particular, we plan to generate vectors to represent, not only words, but also phrases because many biomedical concepts are multiwords. Additionally, the parameters of CRF algorithm will be fine-tuned through cross-validation on the training set for improving the classification results on the test set.

Finally, we would like to investigate the contribution of word embeddings for the relation extraction task, especially, the extraction of DDIs. We will also explore how the DINTO ontology can be used to improve the DDI extraction task. We strongly believe that this ontology could be a valuable resource for the research on Biomedical Information Extraction and would like to encourage the research community to use the DINTO ontology, which is available for research purposes at https://code.google.com/p/dinto/.

[10] http://www.whocc.no/atc/structure_and_principles/

[11] http://www.accessdata.fda.gov/scripts/cder/drugsatfda/

Acknowledgments

This work was supported by TrendMiner project [FP7-ICT287863] and by eGovernAbility-Access project (TIN2014-52665-C2-2-R).

References

Kirill Degtyarenko, Paula De Matos, Marcus Ennis, Janna Hastings, Martin Zbinden, Alan McNaught, Rafael Alcántara, Michael Darsow, Mickaël Guedj, and Michael Ashburner. 2008. Chebi: a database and ontology for chemical entities of biological interest. *Nucleic acids research*, 36(suppl 1):D344–D350.

Zellig S Harris. 1954. Distributional structure. *Word*.

Yongqun He, Sirarat Sarntivijai, Yu Lin, Zuoshuang Xiang, Abra Guo, Shelley Zhang, Desikan Jagannathan, Luca Toldo, Cui Tao, and Barry Smith. 2014. Oae: the ontology of adverse events. *J Biomed Semantics*, 5:29.

María Herrero-Zazo, Isabel Segura-Bedmar, Paloma Martínez, and Thierry Declerck. 2013. The DDI corpus: An annotated corpus with pharmacological substances and drug-drug interactions. *Journal of Biomedical Informatics*, 46(5):914–920.

María Herrero Zazo. 2015. *Semantic Resources in Pharmacovigilance: A Corpus and an Ontology for Drug-Drug Interactions*. Ph.D. thesis, Carlos III University of Madrid, 5.

Martin Krallinger, Florian Leitner, Obdulia Rabal, Miguel Vazquez, Julen Oyarzabal, and Alfonso Valencia. 2015a. Chemdner: The drugs and chemical names extraction challenge. *J Cheminform*, 7(Suppl 1):S1.

Martin Krallinger, Obdulia Rabal, Florian Leitner, Miguel Vazquez, David Salgado, Zhiyong Lu, Robert Leaman, Yanan Lu, Donghong Ji, Daniel M Lowe, et al. 2015b. The chemdner corpus of chemicals and drugs and its annotation principles. *Journal of cheminformatics*, 7(Suppl 1):S2.

Robert Leaman, Chih-Hsuan Wei, and Zhiyong Lu. 2015. tmchem: a high performance approach for chemical named entity recognition and normalization. *Journal of cheminformatics*, 7(supplement 1).

Shengyu Liu, Buzhou Tang, Qingcai Chen, Xiaolong Wang, and Xiaoming Fan. 2015. Feature engineering for drug name recognition in biomedical texts: Feature conjunction and feature selection. *Computational and mathematical methods in medicine*, 2015.

Tomas Mikolov, Kai Chen, Greg Corrado, and Jeffrey Dean. 2013a. Efficient estimation of word representations in vector space. In *ICLR 2013 Workshop Track*.

Tomas Mikolov, Ilya Sutskever, Kai Chen, Greg S Corrado, and Jeff Dean. 2013b. Distributed representations of words and phrases and their compositionality. In *Advances in neural information processing systems*, pages 3111–3119.

Naoaki Okazaki. 2007. Crfsuite: a fast implementation of conditional random fields (crfs).

T. Rocktäschel, M. Weidlich, and U. Leser. 2012. Chemspot: a hybrid system for chemical named entity recognition. *Bioinformatics*, 28(12):1633–1640.

Tim Rocktäschel, Torsten Huber, Michael Weidlich, and Ulf Leser. 2013. Wbi-ner: The impact of domain-specific features on the performance of identifying and classifying mentions of drugs. In *Proceedings of the 7th International Workshop on Semantic Evaluation*, pages 356–363.

Isabel Segura Bedmar, Paloma Martínez, and María Herrero Zazo. 2013. Semeval-2013 task 9: Extraction of drug-drug interactions from biomedical texts. In *SemEval-2013: Semantic Evaluation Exercises Workshop*, pages 341–350. Association for Computational Linguistics.

Isabel Segura-Bedmar, Paloma Martínez, and María Herrero-Zazo. 2014. Lessons learnt from the ddiextraction-2013 shared task. *Journal of biomedical informatics*, 51:152–164.

Richard Socher, John Bauer, Christopher D Manning, and Andrew Y Ng. 2013a. Parsing with compositional vector grammars. In *In Proceedings of the ACL conference*. Citeseer.

Richard Socher, Alex Perelygin, Jean Y Wu, Jason Chuang, Christopher D Manning, Andrew Y Ng, and Christopher Potts. 2013b. Recursive deep models for semantic compositionality over a sentiment treebank. In *Proceedings of the conference on empirical methods in natural language processing (EMNLP)*, volume 1631, page 1642. Citeseer.

Joseph Turian, Lev Ratinov, and Yoshua Bengio. 2010. Word representations: a simple and general method for semi-supervised learning. In *Proceedings of the 48th annual meeting of the association for computational linguistics*, pages 384–394. Association for Computational Linguistics.

David S Wishart, Craig Knox, An Chi Guo, Savita Shrivastava, Murtaza Hassanali, Paul Stothard, Zhan Chang, and Jennifer Woolsey. 2006. Drugbank: a comprehensive resource for in silico drug discovery and exploration. *Nucleic acids research*, 34(suppl 1):D668–D672.

System	Feature set
CRF	standard feature set
CRFD	baseline + DINTO feature
CRFclusterK50Wiki	CRFD's features + word cluster from Word2Vec trained with k=50 on Wikipedia
CRFclusterK50MedLine	CRFD's features + word cluster from Word2Vec trained with k=50 on MedLine
CRFclusterK150Wiki	CRFD's features + word cluster from Word2Vec trained with k=150 on Wikipedia
CRFclusterK150MedLine	CRFD's features + word cluster from Word2Vec trained with k=150 on MedLine
CRFclusterK500Wiki	CRFD's features + word cluster from Word2Vec trained with k=500 on Wikipedia
CRFclusterK50MedLine	CRFD's features + word cluster from Word2Vec trained with k=500 on MedLine
CRFvec50Wiki	CRFD's features + word vectors of dimension 50 from Word2Vec trained on Wikipedia
CRFvec50MedLine	CRFD's features + word vectors of dimension 50 from Word2Vec trained on MedLine
CRFvec100Wiki	CRFD's features + word vectors of dimension 100 from Word2Vec trained on Wikipedia
CRFvec100MedLine	CRFD's features + word vectors of dimension 100 from Word2Vec trained on MedLine
CRFvec200Wiki	CRFD's features + word vectors of dimension 200 from Word2Vec trained on Wikipedia
CRFvec200MedLine	CRFD's features + word vectors of dimension 200 from Word2Vec trained on MedLine

Table 2: List of experiments.

		Exact criterion			Strict criterion		
		P	R	F1	P	R	F1
DDI-DrugBank	WBI	0.90	0.89	0.90	0.88	0.87	0.87
	CRF	0.70	0.85	0.77	0.69	0.82	**0.75**
	CRFD	0.72	0.84	0.77	0.68	0.81	0.74
	CRFclusterK50Wiki	0.72	0.89	0.79	0.68	0.83	**0.75**
	CRFclusterK150Wiki	0.73	0.89	**0.80**	0.68	0.83	0.74
	CRFclusterK500Wiki	0.72	0.89	**0.80**	0.68	0.83	0.74
	CRFclusterK50MedLine	0.72	0.86	0.79	0.69	0.82	**0.75**
	CRFclusterK150MedLine	0.72	0.86	0.79	0.68	0.82	0.74
	CRFclusterK500MedLine	0.72	0.86	0.79	0.69	0.82	**0.75**
	CRFvec50Wiki	0.71	0.84	0.77	0.69	0.81	0.74
	CRFvec100Wiki	0.72	0.84	0.77	0.69	0.81	0.74
	CRFvec200Wiki	0.72	0.85	0.78	0.68	0.80	0.74
	CRFvec50MedLine	0.72	0.84	0.78	0.69	0.82	**0.75**
	CRFvec100MedLine	0.73	0.86	0.79	0.68	0.81	0.74
	CRFvec200MedLine	0.73	0.85	0.79	0.68	0.80	0.74
DDI-MedLine	WBI	0.81	0.74	0.77	0.61	0.56	0.58
	CRF	0.69	0.54	0.61	0.62	0.44	0.52
	CRFD	0.79	0.57	0.66	0.70	0.47	0.56
	CRFclusterK50Wiki	0.74	0.63	**0.68**	0.66	0.48	0.56
	CRFclusterK150Wiki	0.73	0.63	**0.68**	0.67	0.49	**0.57**
	CRFclusterK500Wiki	0.72	0.64	**0.68**	0.65	0.51	**0.57**
	CRFclusterK50MedLine	0.74	0.59	0.66	0.64	0.46	0.53
	CRFclusterK150MedLine	0.75	0.63	**0.68**	0.66	0.49	0.56
	CRFclusterK500MedLine	0.73	0.62	0.67	0.67	0.49	**0.57**
	CRFvec50Wiki	0.77	0.57	0.66	0.68	0.47	0.56
	CRFvec100Wiki	0.78	0.56	0.66	0.66	0.46	0.54
	CRFvec200Wiki	0.77	0.57	0.66	0.68	0.46	0.55
	CRFvec50MedLine	0.79	0.57	0.66	0.66	0.45	0.54
	CRFvec100MedLine	0.81	0.57	0.66	0.69	0.46	0.55
	CRFvec200MedLine	0.78	0.57	0.66	0.68	0.46	0.55

Table 3: Experimental results.

Creating a rule based system for text mining of Norwegian breast cancer pathology reports

Rebecka Weegar and Hercules Dalianis
Department of Computer and Systems Sciences
Stockholm University
P.O. Box 7003
164 07 Kista
Sweden
{rebeckaw,hercules}@dsv.su.se

Abstract

National cancer registries collect cancer related information from multiple sources and make it available for research. Part of this information originates from pathology reports, and in this pre-study the possibility of a system for automatic extraction of information from Norwegian pathology reports is investigated. A set of 40 pathology reports describing breast cancer tissue samples has been used to develop a rule based system for information extraction. To validate the performance of this system its output has been compared to the data produced by experts doing manual encoding of the same pathology reports. On average, a precision of 80%, a recall of 98% and an F-score of 86% has been achieved, showing that such a system is indeed feasible.

1 Introduction

Cancer is a common cause for death worldwide, with about 14 million new cases each year (World Health Organization, 2014). In the Nordic countries it is mandatory to report each incidence of cancer to national registries and in Norway, the reported data is handled by the Cancer Registry of Norway, (Kreftregisteret i Oslo). The registry has as its main functions to monitor the cancer prevalence in Norway by collecting data on all incidences of cancer, and also to make this data available for research (Ministry of Health and Care Services, 2001). In 2013, there were about 30,000 new cases of cancer in Norway, the most common cancer type for women being breast cancer with 3,220 new cases, and the most common type for men being prostate cancer with 4,836 new cases (Cancer Registry of Norway, 2015).

Part of the data that the Cancer Registry of Norway handles originates from pathology reports. A pathology report is written by a pathologist examining a tissue sample from a patient with known or suspected cancer and the report contains a number of test results, measurements and descriptions of the sample.

The National Cancer Registry of Norway receives about 180,000 pathology reports each year and 25 full time expert coders transfer data from the free text reports to a database via an XML template. The manual encoding of the pathology reports requires special knowledge for each cancer type and the transferal is a complicated and time consuming task where the coders have to read and interpret the content of each report.

There is therefore a need of a system capable of automatic information extraction. The system should be able to accurately extract the relevant fields for each type of cancer.

2 Related research

Several studies have been performed on information extraction in the domain of pathology reports with the aim to structure their contents (Spasic et al., 2014). Rule based systems and machine learning systems are both used, and in some cases in combination. Coden et al. (2009) built a model called Cancer Disease Knowledge Representation Model, which has nine classes including anatomical site, histology, and metastatic tumor. Evaluation found that recall was between 76% and 100% and precision was between 72% and 100% for all classes except metastatic tumor where both precision and recall were lower.

Kavuluru et al. (2013) extracted the anatomical location of neoplasms from pathology reports describing several types of cancers. They achieved an average micro F-score of 90% and an average macro F-score of 72%.

Xu et al. (2004), used the MedLee system to analyze breast cancer pathology reports and had a performance for tabular findings of 95.8% sensi-

Proceedings of the Sixth International Workshop on Health Text Mining and Information Analysis (Louhi), pages 73–78,
Lisbon, Portugal, 17 September 2015.

tivity (recall) and 95.4% precision. For narrative text these numbers became lower with 90.6% sensitivity (recall) and 91.6% precision.

Currie et al. (2006), constructed a rule based system to extract concepts from 5,826 breast cancer and 2,838 prostate cancer pathology reports. The authors obtained around 90-95% accuracy for most of the 80 extracted fields, using domain experts for the evaluation.

Ou and Patrick (2014) studied pathology reports concerning primary cutaneous melanomas. They used both rule and machine learning based approaches. Their system was evaluated on 97 reports and they obtained an average F-score of 85% on identifying 28 different concepts including diagnosis, size and laterality and tumor thickness.

Schadow and McDonald (2003), used 275 surgical pathology reports in their experiments. Their regular expression based parser identified around 90% of the codings correctly.

McCowan et al. (2007), Nguyen et al. (2010) and Martínez et al. (2014) use text mining to perform cancer classification according to the TNM-scale (Tumor Node Metastases) (Wittekind et al., 2014).

McCowan et al. (2007), trained on 710 pathology reports for lung cancer using the SVM algorithm and evaluated on 179 reports. They obtained an accuracy of 74% for tumor staging and 87% for node Staging. Nguyen et al. (2010), developed a rule based staging system for lung cancer using 100 lung cancer pathology reports and evaluated it on 718 reports. The authors obtained an accuracy of 72%, 78%, and 94% for tumor, node, and metastases staging, respectively. Martínez et al. (2014), obtained F-scores of 81%, 85%, and 94% for staging tumor, node, and metastases respectively for colorectal cancer pathology reports. The authors used 200 pathology reports for training and evaluation.

Although closely related and relevant to this study, these studies are all performed on pathology reports in English; therefore the systems are not directly applicable to the Norwegian reports. To the best of our knowledge, only one study of information extraction from Norwegian pathology reports exists. Singh et al. (2015) used 25 pathology reports related to prostate cancer as input data. They used SAS Institute software to extract fields and they report a percentage of correctly extracted fields of 76% for number of biopsies, 24% for number of biopsies containing tumor tissue, and 100% for Gleason score. The study focuses on system development and it is not clear if they divided the data into a development set and a test set.

3 Material and methods

The Cancer Registry of Norway has selected a set of 40 pathology reports in XML-format for this pre-study. The reports have been manually de-identified by the registry and fields identifying individual patients have been removed.

The content of a pathology report depends on the procedure that produced the tissue sample. For this study the selected report types are mastectomy, where the whole breast is removed, and breast-conserving surgery, where a smaller piece is removed. Figure 1 shows an example of a portion of free text from a pathology report. It describes a tissue sample with invasive ductal carcinoma and ductal carcinoma in situ, and the measured margins around both the invasive carcinoma and the carcinoma in situ. It also mentions the percentage of estrogen receptor positive cells, progesterone positive cells and the presence of the Ki67 marker.

A program for extracting free text fields and encoded data fields from the XML-files has been written, and the input text has been divided into tokens using a custom program. A token corresponds to a unit of text, which can be a word, a number or punctuation sign, percentage sign etc. The number of tokens in the reports is ranging from 107 to 1,203 tokens with a median of 531 tokens. There are 22,670 tokens in total in the input data.

Input text and corresponding encoding

The pathology reports used in this study consist of two parts, the free text part written by a pathologist and the encoding of the same report performed by an expert coder. Each encoded field and its possible values are described in the internal requirements defined by the registry (Kreftregisteret, 2014). The requirements do, however, not say anything about how the pathologists should write their reports; the input text is therefore not as well defined as the encoded parts of the reports.

The free text contains both macroscopic and microscopic descriptions of the tissue sample. The descriptions can include test results, size measure-

```
Mammaresektat (ve.  side) med
infiltrerende duktalt karsinom,
histologisk grad 3
Tumordiameter 15 mm
Lavgradig DCIS med utstrekning 4 mm i
kranial retning fra tumor
Frie reseksjonsrender for infiltrerende
tumor (3 mm kranialt)
Lavgradig DCIS under 2 mm fra kraniale
reseksjonsrand

ER: ca 65 % av cellene positive
PGR: negativ
Ki-67:  Hot-spot 23% positive celler.
Cold spot 8%.  Gjennomsnitt 15%
HER-2:  negativ
Tidl.  BU 13:

3 sentinelle lymfeknuter uten påviste
patologiske forandringer
```

Figure 1: Extract from the free text part of an anonymised breast cancer report in Norwegian, but the data in the figure is made up and can not be linked to any individual.

ments, the type of cancer and the possible degree of hormone receptors. Other reported findings are pre-cancers and metastases in lymph nodes.

Some of the values are explicitly stated in the text as for example tumor size in Figure 1 *Tumordiameter 15 mm* (Tumor diameter 15 mm). Other values are implicit and need to be inferred from the text.

An example of this is the pT-values. They are a kind of staging information for tumors, and in the case of breast cancer the pT-value is based on the size of the tumor and what tissues the tumor is growing in (Naume, 2015). The pT value is not explicitly stated in the text, so the human or machine encoder needs to evaluate several parts of the text to determine the value of such a field.

A small portion of values appears in the same form in the input text as in the encoding, but many of the values are translated into one of a set of predefined values. For example, estrogen receptors are reported in numerical values in the text, as in Figure 1 *ER: ca 65 % av cellene positive*. This percentage value is discretized to one of six possible values when coded.

In total there can be 83 encoded fields for a single report. There are 47 different field types and 18 of the field types can be repeated up to three times depending on the number of tumors present in the tissue sample. A majority of the fields are mandatory to encode, but an option such as *not performed* is often available.

The distribution of textual and encoded fields is presented in Table 1. The implicit type is most common in the input texts and the discretized type is most common in the encodings. There is an average of 5 different values for the discretized fields.

	Encoding type	Continuous	Discrete	True/False	*Total*
Textual type	Continuous	19%	4%		**23%**
	Discrete		11%		**11%**
	Implicit		17%	30%	**47%**
	Cont./Impl.		19%		**19%**
	Total	**19%**	**51%**	**30%**	**100%**

Table 1: The 47 encoded values sorted by type, the Cont./Impl. category contains the values that are present either as continuous or implicit values in the input texts.

4 A rule based approach for information extraction

The available pathology reports have been divided into a development set of 30 reports and a test set of ten reports. The encoding of the reports has been used for evaluation and there has not been any additional manual annotation of the free text.

The developed system is based on the idea that specific fields are identified by their form and context. There are, for example, a number of fields in the reports that are reported in the form of percentages and it is possible to distinguish them by looking at characteristic tokens appearing before and after them.

Each field therefore gets assigned one or more Regex-style rules and two optional lists containing sequences of tokens. The first list holds sequences associated with the field and appearing before it, and the second contains sequences appearing after the field. The content of the context lists was created by manual inspection of the pathology reports in the development set.

One example of a field in the reports is the Ki67 hot spot value. It is often explicitly stated in the text in the form of a percentage. Therefore, the token % has been put in the after-list, and the token sequences selected for the before-list were *hot spot, hotspot, hot spotområde, ki* - and *ki67*. A program was then used to search each sentence in the data for these tokens and a regular expression was used to extract any numerical values found between them.

An automatic approach for creating the context lists has also been tested. Each unigram, bigram, trigram and 4-gram appearing in the development set was evaluated in three steps; scoring, sorting and selecting. In the first step the individual n-gram was scored using F-scores according to its ability to extract the correct values for an investigated field. In the second step, the n-grams were sorted in descending order according to this score and in the final step a set of n-grams were selected. The selection was performed by taking each n-gram in order and putting it into the context list. If the adding of the n-gram increased the total F-score for the field, the n-gram was kept in the list.

5 Results

The system has been evaluated against the manual encoding using precision, recall and F-score. The results are presented in Table 2. The fields Sentinel Nodes and the Axillary Nodes can have two possible values, *performed* and *not performed*. The field Tumor size is encoded in millimeters and therefore has many possible values. Ki67 is a protein indicating the growing rate of tumors and the two different Ki67 fields are encoded in percent. The hormone receptors for estrogen and progesterone are also reported in percent, but encoded into five and six different values, respectively. It is also possible for these values to be encoded as *not stated* if they are not present in the reports. The pT-value can be encoded as 18 different values depending on the size of the tumor, the type of cancer and where the cancer grows.

Data set	Development set			Test set		
Field	P	R	F	P	R	F
Sentinel Nodes	83	100	91	60	100	75
Axillary Nodes	93	100	97	90	100	95
Tumor Size	77	91	83	78	100	88
Histological grade	96	96	96	100	100	100
Estrogen	77	100	87	70	100	82
Progesterone	83	100	91	70	100	82
Receptors N.R.	93	100	97	90	100	95
HotSpot Ki67	93	100	97	90	100	95
Avg. Ki67	39	100	56	100	75	86
pT	80	100	88	50	100	67
Average all	82	99	88	80	98	86

Table 2: The precision (P), recall (R) and F-score (F) achieved on the test and development data in percent. N.R. stands for not reported.

The automatic creation of context lists was tested on four of the fields, histological grade, Ki67 hot spot value, Ki67 average value and tumor size; see Table 3. The automatically created context list for tokens appearing before the Ki67 hot spot value contained *hot*, *ki67*, - and *hotspot*.

	Development set			Test set		
Field	P	R	F	P	R	F
Hist. grade	96	96	96	50	100	67
Tumor size	81	96	88	88	88	88
HotSpot Ki67	93	100	97	90	100	95
Avg. Ki67	39	100	56	100	75	86
Avg. automat.	77	98	84	82	91	84
Avg. manual	76	97	83	92	94	92

Table 3: The achieved precision (P), recall (R) and F-score (F) in percent when using the automatically created context lists. The last row shows the average scores on the same four fields when using the manual approach.

6 Conclusions and Future work

In this pre-study, the possibility a system for extracting information from pathology reports written in Norwegian has been investigated.

A number of different encoding types have been identified in the data. This suggests the need for a number of approaches for successful information extraction. One main difficulty is to determine whether a value is actually present in the report, since not all tests are preformed on all tissue samples. Here, text classification could be imagined as a useful technique. Several of the fields in the reports are explicitly stated in a limited number of possible ways. In these cases, a rule based approach as the one presented here could perform well. There is also a category of values where the encoding is more complicated. This is the case when several parts of the input text needs to be interpreted to find the correct encoding, here different machine learning techniques should be investigated. An overview of the future system is shown in Figure 2.

The manually created context lists gave a better performance than the automatically created context lists. This can be explained by the fact that a human can imagine similar contexts to the ones found in the development data and add those to the context lists. The automatic creation could, however, be useful when using more data and when expanding to other types of cancers since it requires no or little manual inspection of the input texts.

The validity of the presented precision, recall

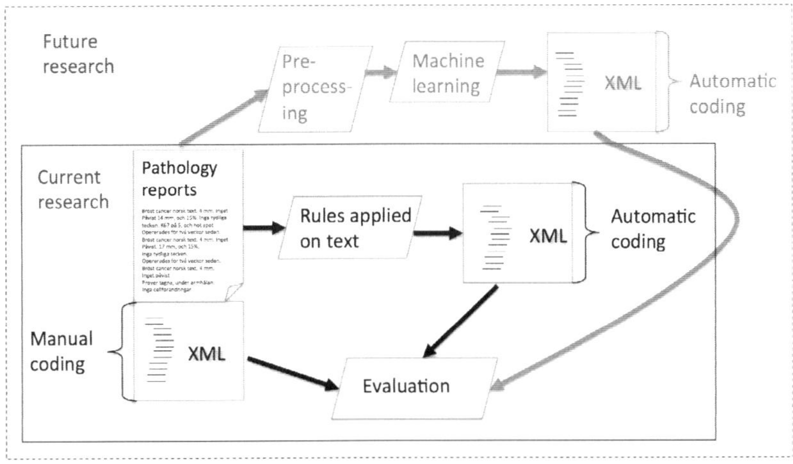

Figure 2: The pathology mining system

and F-scores for the information extraction can not be considered as very high, as too little data has been used. To make any robust claims about the performance of a future system, more test data is needed, and to properly develop the system more development data is also crucial. Ideally the performance of this system should be compared to an inter-annotator-agreement measure for the expert coders. However, the achieved results are promising and show that this system should be further developed and that a well functioning system is feasible.

Acknowledgements

The authors would like to thank Dr. Jan Nygård at Kreftregisteret Oslo, for his assistance with data access and Kjersti Østby for providing domain knowledge.

This work was supported by the Nordic Information for Action eScience Center (NIASC); a Nordic Center of Excellence financed by NordForsk (Project number 62721).

References

Cancer Registry of Norway. 2015. Cancer in Norway 2013 - Cancer incidence, mortality, survival and prevalence in Norway.

Anni Coden, Guergana Savova, Igor Sominsky, Michael Tanenblatt, James Masanz, Karin Schuler, James Cooper, Wei Guan, and Piet C. de Groen. 2009. Automatically extracting cancer disease characteristics from pathology reports into a disease knowledge representation model. *J. of Biomedical Informatics*, 42(5):937–949, October.

Anne-Marie Currie, Travis Fricke, Agnes Gawne, Ric Johnston, John Liu, and Barbara Stein. 2006. Automated extraction of free-text from pathology reports. In *AMIA*.

Ramakanth Kavuluru, Isaac Hands, Durbin Eric B., and Lisa Witt. 2013. Automatic extraction of icd-o-3 primary sites from cancer pathology reports. In *AMIA Jt Summits Transl Sci Proc*.

Oslo Kreftregisteret. 2014. *Requirements specification for breast cancer reporting to the Cancer Registry of Norway, Internal document*. The Cancer Registry of Norway.

David Martínez, Graham Pitson, Andrew MacKinlay, and Lawrence Cavedon. 2014. Cross-hospital portability of information extraction of cancer staging information. *Artificial Intelligence in Medicine*, 62(1):11–21.

Iain A McCowan, Darren C Moore, Anthony N Nguyen, Rayleen V Bowman, Belinda E Clarke, Edwina E Duhig, and Mary-Jane Fry. 2007. Collection of cancer stage data by classifying free-text medical reports. *Journal of the American Medical Informatics Association*, 14(6):736–745.

Ministry of Health and Care Services. 2001. Regulations on the collection and processing of personal health data in the Cancer Registry of Norway (Cancer Registry Regulations).

Bjørn Naume. 2015. Stadier ved brystkreft, http://www.oncolex.no/bryst/bakgrunn/stadier.

Anthony N Nguyen, Michael J Lawley, David P Hansen, Rayleen V Bowman, Belinda E Clarke, Edwina E Duhig, and Shoni Colquist. 2010. Symbolic rule-based classification of lung cancer stages from free-text pathology reports. *Journal of the American Medical Informatics Association*, 17(4):440–445.

Ying Ou and Jon Patrick. 2014. Automatic population of structured reports from narrative pathology reports. In Jim Warren and Kathleen Gray, editors, *Seventh Australasian Workshop on Health Informatics and Knowledge Management (HIKM 2014)*, volume 153 of *CRPIT*, pages 41–50, Auckland, New Zealand. ACS.

Gunther Schadow and Clement J McDonald. 2003. Extracting structured information from free text pathology reports. In *AMIA Annual Symposium Proceedings*.

Harneet Singh, Mathias Knudsen Sollie, Emil Orholm Solhøi, and Fredrik Sverre Syberg. 2015. Information Extraction: The Case of Kreftregisteret, (In Norwegian). Bachelor thesis, Westerdals Oslo ACT.

Irena Spasic, Jacqueline Livsey, John A. Keane, and Goran Nenadic. 2014. Text mining of cancer-related information: Review of current status and future directions. *I. J. Medical Informatics*, 83(9):605–623.

Christian Wittekind, Hisao Asamura, and Leslie H Sobin (eds). 2014. *TNM Atlas. Ilustrated Guide of the TNM Classification of Malignant Tumours*. Sixth edition, Wiley Blackwell.

World Health Organization. 2014. World cancer report 2014.

Hua Xu, Kristin Anderson, Victor R Grann, and Carol Friedman. 2004. Facilitating cancer research using natural language processing of pathology reports. *Medinfo*, 11(Pt 1):565–72.

Parser Adaptation to the Biomedical Domain without Re-Training

Jeff Mitchell
School of Informatics
University of Edinburgh
Edinburgh, EH8 9AB, UK
jeff.mitchell@ed.ac.uk

Mark Steedman
School of Informatics
University of Edinburgh
Edinburgh, EH8 9AB, UK
steedman@inf.ed.ac.uk

Abstract

We present a distributional approach to the problem of inducing parameters for unseen words in probabilistic parsers. Our KNN-based algorithm uses distributional similarity over an unlabelled corpus to match unseen words to the most similar seen words, and can induce parameters for those unseen words without retraining the parser. We apply this to domain adaptation for three different parsers that employ fine-grained syntactic categories, which allows us to focus on modifying the lexicon, while leaving the structure of the parser itself intact. We demonstrate uplifts for dependency recovery of 2%-6% on novel vocabulary in biomedical text.

1 Introduction

Parsing is an important component in many NLP applications. Shallower analyses may allow the discovery of local relations, but to handle the full complexity of speech and text requires knowledge of the hierarchical structures that parsers are designed to uncover. This is particularly true of long range dependencies such as that between *activities* and *decreased* in *the specific synthetic activities of electrophoretically purified myosin heavy chain decreased*. Such dependencies have proven to be useful features in many text mining and knowledge extraction applications, for example identifying biomarkers in the biomedical literature (Seoud and Mabrouk, 2013) or extracting family history from clinical text (Lewis et al., 2011).

Correctly identifying the dependencies within a string of words is generally based on finding the most probable structure over them, and this in turn requires knowing what sort of relations each word is likely to enter into. Unfortunately, gold standard training data, annotated with these syntactic relations, is generally in short supply. The vocabulary for which we have explicitly seen examples of the type of dependencies each word supports is therefore typically small and performance on real data is often degraded in handling out-of-vocabulary items.

Although the Penn Treebank has been a vital tool in the development and evaluation of parsing technology, providing a standard dataset for comparison of parsers, practical application of these techniques usually requires adaptation to new domains. Rimell and Clark (2009), for example, examine the adaptation of a WSJ-trained CCG parser to the biomedical domain. The divergence between these two domains, news and biology, is manifest in terms of both vocabulary and also stylistic differences in the prevalence of various syntactic structures. For example, biomedical writing eschews personal pronouns and tolerates long sequences of noun modifiers, whereas the style of news articles tends to reverse these preferences. Rimell and Clark's (2009) approach to adapting to these differences is based on retraining elements of the model using biomedical texts which have been hand-tagged with gold-standard tags. While this is undoubtedly effective, achieving an overall improvement of F-score of over 5%, it requires a considerable commitment of skilled resources to manually annotate a substantial corpus with the linguistically correct tags.

Here, we consider a distributional approach to domain adaptation using the information about syntactic structure that is implicit in raw text. We estimate parameters for unseen words using a KNN approach that matches them to the nearest seen words and averages over their parameters. We explore a number of different approaches to measuring distributional similarity and find that vectors based on counts of occurrence within ngram contexts give the best results. Bag-of-word approaches and neural embeddings, which have worked well for semantic tasks, do not appear to

Proceedings of the Sixth International Workshop on Health Text Mining and Information Analysis (Louhi), pages 79–89,
Lisbon, Portugal, 17 September 2015.

capture the information about syntactic similarity that this task requires.

Our use of ngram contexts is inspired by psycholinguistic research into the acquisition of syntactic categories. Cartwright and Brent (1997), for example, consider how children might use a word's distribution across a range of templates, such as $\langle the\ XXX\ is\ good \rangle$, to infer its syntactic properties. They show, in simulations, that such distributional information can be used to infer syntactic categories from child-directed speech. Mintz (2003) analyses distributions over a simpler type of template, which he calls a frequent frame, consisting of a pair of common lexical items flanking a word of interest, e.g. $\langle you\ XXX\ it \rangle$ or $\langle the\ XXX\ is \rangle$. In addition to showing how such distributional information can be used to induce categories, he also discusses the evidence that adults and children are sensitive to these frames. Redington et al. (1998) consider even simpler contexts, based simply on bigram colocations, e.g. $\langle the\ XXX \rangle$. Pinker (Pinker, 1987), on the other hand, has long contested the possibility of using such distributional information to acquire valid grammatical categories, and proposes instead that grammatical categories are bootstrapped using semantic knowledge.

While the patterns and templates described above can be used to characterise a word's behaviour in terms of concrete occurrences in specific contexts, neural networks have recently become popular as a means to create more abstract representations. In this case, as the network adapts to the data, representations are learned that embed discrete inputs in a continuous space defined by its internal states. Researchers have been interested in the nature of such internal representations for some time (e.g., Small et al., 1995; Joanisse and Seidenberg, 1999). However, it has now become practical to induce such embeddings from large quantities of text and employ them in linguistic applications. For example, Tsuboi (2014) and Collobert et al. (2011) apply neural representations to POS tagging, and this suggests that at least some useful information about the syntax of unseen words might be gained from this source.

While POS tags can provide a coarse-grained description of words' syntactic behaviour, accurate parsing typically requires finer-grained detail. We can distinguish between two approaches, which may be combined, to specifying this additional level of detail. The first approach simply makes use of finer-grained syntactic categories, either instead of or in addition to POS tags (Steedman, 2000; Klein and Manning, 2003b; Petrov et al., 2006). These categories can then determine the missing information about the dependencies a word will take part in, such as whether a verb is intransitive or whether it takes prepositional arguments. The second approach instead increases the granularity of the production rules, by conditioning the probabilities on the heads of the phrases involved (Charniak, 2001; Collins, 2003). In this way, words are associated with probabilities for the structure of phrases that they head, determining, for example, the types of object that a verb phrase expands into.

Although the two approaches are compatible, a significant difference makes the former more conducive to our purposes. Enhancing the granularity of the syntactic categories results in a much richer lexicon containing more information about how words behave syntactically. In principle, this should lead to an enlargement of the lexicon having a greater impact on performance by itself. In the latter approach, of lexicalising the production rules, expanding the vocabulary of the parser may be much more complicated, requiring modifications throughout the model. In contrast our approach simply adds new entries to the lexicon without the need to retrain the parser. In fact, our approach does not even require full sentences and can be applied to an unlabelled corpus of ngram counts.

Our KNN approach and the three parsers we modify are described in Sections 2 and 3 respectively. We then use a biomedical dependency recovery task, specified in Section 4, to evaluate the performance of the modified parsers, as reported in Section 5.

2 Approach

Our approach is based on the assumption that words with similar syntactic properties should have similar distributional characteristics. We evaluate both neural embeddings and also raw context frequencies as the basis for measuring distributional similarity. These context vectors have components which correspond to occurrences within a corpus of raw biomedical text and we employ both SENNA (Collobert et al., 2011) and Skip-gram (Mikolov et al., 2013) embeddings. In all cases, we induce parameters for unseen words by averaging the parameters from the k nearest neighbours seen in the training data.

2.1 Context Vectors

Distributional similarity is here based on comparing vectors that are constructed from raw context counts. We considered two approaches to defining these contexts: ngrams and bags-of-word (BOW).

The ngram approach counts occurrences in 2gram, 3gram and 4gram contexts that are intended to emphasise syntactic - as opposed to semantic - characteristics, following the structure of templates and frames proposed by e.g. Cartwright and Brent (1997), Mintz (2003) and Redington et al. (1998). Thus our 2gram contexts have two forms that distinguish occurrence on the left from occurrence on the right: $\langle left\_token\ XXX \rangle$ and $\langle XXX\ right\_token \rangle$. The 3gram contexts are equivalent to Mintz's (2003) frequent frames: $\langle left\_token\ XXX\ right\_token \rangle$. And the 4gram contexts extend this frame to the right, mimicking the form of templates described by Brent (1991) and Cartwright and Brent (1997): $\langle left\_token\ XXX\ right\_token_1\ right\_token_2 \rangle$.

The BOW approach ignores the sequential information contained in the ngram contexts and relies instead on counts of individual words that occur anywhere in 5 word-windows each side of a target word.

In each case, we built distributional vectors using the most common of these contexts, with vector components based on a ratio of probabilities.

$$v_i = \frac{p(c_i|w_t)}{p(c_i)} = \frac{freq_{i,t} \cdot freq_{total}}{freq_i \cdot freq_t} \quad (1)$$

where c_i is the ith context, w_t is the target word, $freq_{i,t}$ is the count of the number of times w_t occurs in context c_i, $freq_i$ is the overall count of the number of times context c_i occurs with all words, $freq_t$ is the overall count for w_t in all contexts and $freq_{total}$ is the total count for all words in all contexts. Target words with $freq_t < 10$ were discarded as containing too little useful information.

The distance between two vectors, **u** and **v**, was measured in terms of the city block metric:

$$dist(\mathbf{u}, \mathbf{v}) = \sum_i |u_i - v_i| \quad (2)$$

This appeared to work more effectively on sparse vectors than the more usual cosine metric. We built these representations on a corpus of 1.2 billion words of titles and abstracts from the Medline database.

2.2 SENNA

Collobert et al. (2011) trained a neural net language model on a snapshot of the English Wikipedia ($\approx$ 631M words) and published the feature vectors[1] induced for each word in the first hidden layer of the network. They showed that these embeddings are useful in enhancing the performance of a number of tasks, including POS tagging and semantic role labelling. Using these representations as features, Bansal et al. (2014) obtained improvements in dependency recovery in the MST Parser (McDonald and Pereira, 2006).

Andreas and Klein (2014) also used these embeddings on a number of tasks, including an attempt to expand the vocabulary of the Berkeley Parser by matching unseen words to the nearest word already in the lexicon. However, instead of inducing parameters for the new vocabulary they simply replaced unseen words with their seen matches in the input. Unfortunately they did not find a reliable benefit from this approach.

Like the context vectors described above, the SENNA representations were derived from large quantities of raw text and reflect the distributional behaviour of words in that data. However, unlike our context vectors, which have components derived from explicit distributional contexts, the components of their neural embeddings are abstract dimensions whose values derive from the optimization of a particular mathematical model. In this case the form of this model was based on distinguishing between real 11-word phrases drawn from the unlabelled corpus and an incorrect phrase which had the central word replaced with a randomly chosen item. The model tries to maximise the difference between these two phrases in terms of scores which are a nonlinear function of the vectors representing the words they contain.

Training involved stochastic gradient ascent optimisation of an objective function based on a ranking criterion for the two phrase scores, and resulted in each word within a 100,000 word vocabulary being assigned a vector representation. The published embeddings are of dimension 50 and we measured the similarity of these vectors in terms of the cosine measure:

[1] http://ronan.collobert.com/senna/

$$dist(\mathbf{u}, \mathbf{v}) = \frac{\mathbf{u} \cdot \mathbf{v}}{|\mathbf{u}||\mathbf{v}|} \quad (3)$$

2.3 Skip-gram

Like the SENNA model, the Skip-gram model (Mikolov et al., 2013) is trained to differentiate between the correct central word of a phrase and a random replacement, which they refer to as negative sampling. Unlike SENNA, however, the Skip-gram model tries to make this prediction using only a single one of the surrounding words at a time and ignores the ordering of those words, i.e. taking a bag-of-words approach to context.

The published 300-dimensional vectors[2] were trained on 100B words of Google News text using stochastic gradient ascent, and cover a vocabulary of 3M words. We also retrained the same 300-dimensional model on our 1.2 billion word unlabelled biomedical corpus, giving a vocabulary of around 1M words. In both cases, we measured similarity using the cosine metric, Equation 3.

2.4 KNN Parameter Induction

Our approach to inducing parser parameters for unseen words is a form of k-nearest-neighbor induction.[3] Specifically, we constructed parameters for unseen words by finding the most similar words in the lexicon, using the distributional measures described above, and then averaging over their existing parameters in the parsing model. We did this for each parser, varying the dimensions of the context vectors, and the number of nearest neighbours to find the optimal model. To ensure that the parameters that we average over are well-estimated and reliable, we only consider words that appear more than a hundred times in the Penn Treebank when finding the nearest neighbours.

3 The Parsers

We extend the vocabulary of three parsers, all of which make use of fine-grained lexical categories. The first of these parsers induces sub-categories beneath the level of POS-tags during training while the other two require hand-annotation of the categories in the training data. In all cases, we modify the parser merely by inserting new items, along with their tag parameters, into the lexicon while leaving the rule probabilities in the rest of the parser unchanged. Sections 3.1, 3.2 and 3.3 outline these parsers, focusing particularly on the contents of the lexicon which our methods modify as decribed in Section 2.

3.1 The Berkeley Parser

While an unlexicalized parser that uses syntactic categories based solely on the symbols found in the Penn Treebank will generally perform poorly, a number of results show that refining these categories can substantially improve performance. Klein and Manning (2003b), for example, show that the performance of an unlexicalised model can be substantially improved by splitting the existing symbols down into finer categories. Their subcategorizations were developed by hand based on linguistic intuitions and a careful error analysis. The Berkeley Parser[4] (Petrov et al., 2006), in contrast, is based on a method for automatically finding useful subcategorizations during training by splitting and merging the original nodes.

The model is an unlexicalized generative PCFG, but the granularity of the terminal and non-terminal categories found in training give it a much greater sensitivity to the syntactic behaviour of words and phrases than is possible using standard POS tags. The lexicon specifies each word's association to the terminal categories, and the rest of the parser is entirely unlexicalized. Parsing is complicated by the large number of syntactic categories which threaten to make standard techniques infeasible, due to the size of the search space and also even just the amount of memory required to hold the chart. However, the hierarchical structure resulting from the split-merge process enables a form of coarse to fine pruning that makes the problem tractable (Petrov and Klein, 2007). Training is based on the EM algorithm along with 6 cycles of splitting each symbol into two and remerging the 50% of sub-symbols carrying the least information. Output from the Berkeley Parser consists of trees labelled with the original Penn Treebank symbols, and we use the EnglishGrammat-

[2] https://code.google.com/p/word2vec/

[3] We also evaluated Support Vector Regression as a means of inducing parameters, but we found it to be less effective. Although the characteristics of SVMs do in general make them powerful modelling tools, this particular task required us to use one SVM model for every parameter type to be induced (e.g. $\approx$ 400 CCG categories). In fact, the requirement to optimise the C and gamma hyper-parameters resulted in evaluation of about 100 models per parameter (e.g. $\approx$ 40,000 models). In contrast, the KNN approach induces all the parameters in one single model, producing a much more constrained problem, which probably contributes to its superior generalisation in this case.

[4] https://code.google.com/p/berkeleyparser/

icalStructure class from the Stanford Parser[5] to convert the trees to Stanford-style dependencies. Out-of-vocabulary items are handled by a process that uses orthography and sentence position to estimate probabilities for unseen words.

Expanding the lexicon of this model using our KNN method is complicated by the fact that it is generative, so that inserting new vocabulary with non-zero probabilities requires adjusting the probabilities of everything else in the lexicon to maintain normalization. Since the parser uses a cutoff of a word count of 100 or lower to determine whether word given tag probabilities are smoothed, we assigned all new vocabulary a count of 101, and partitioned this count according to the induced tag and sub-tag probabilities. In fact, our attempts to use KNN to induce probabilities over the sub-categories below the level of POS tags were fruitless, producing worse results than the original model in all experiments. Thus, we resorted to using the KNN approach to induce POS level probabilities and then basing the lower level probabilities on a 50-50 interpolation of a general profile for each POS tag and the probabilities assigned by the OOV process.

3.2 C&C

Whereas the Berkeley Parser automatically induces a set of fine-grained categories during training in an attempt to maximize parsing performance, the categories of CCG (Steedman, 2000) have been linguistically designed to represent the dependencies that words will support. In particular, they have a close correspondence to the functional types of lambda calculus representations. So, for example, an intransitive verb has the CCG category $S\backslash NP$, which can be interpreted as identifying this as a syntactic structure that takes a noun phrase to its left (represented by $\backslash NP$) to produce a sentence (represented by S). In other words, it is a function from entities of type NP to type S. In comparison, a transitive verb has the type $(S\backslash NP)/NP$, which describes structure that takes a noun phrase to its right ($/NP$) to produce a structure equivalent to an intranstive verb ($S\backslash NP$), which is itself a category looking for an NP to its left to produce a sentence. Thus, the transitive verb category is a function from two NPs - one to the right and one to the left - to an entity of type S.

[5] http://nlp.stanford.edu/software/lex-parser.shtml

The C&C parser[6] (Curran et al., 2007) is a discriminative parser, which has been trained on CCGbank (Hockenmaier and Steedman, 2007), a translation of the Penn Treebank into the CCG formalism. Roughly, the parser can be split into three modules: a POS-tagger, a super-tagger and the parser itself. The POS-tagger assigns fixed POS tags to the text to be parsed, based on a window of five words centred on the word to be tagged. The super-tagger takes these POS tags and words as input and, using the same five token window, passes CCG tags to the parser. The parser in turn tries to build a derivation from the CCG tags it has been given, but can request a re-analysis from the super-tagger if this fails.

Each module uses a log-linear model to predict which structures, ω, are most likely given the input, S:

$$p(\omega|S) = \frac{e^{\sum_i \lambda_i f_i(\omega)}}{Z_S} \qquad (4)$$

where the f_i are a set of features, the λ_i are feature weights and Z_S is a normalising constant.

Here we only consider modifying the POS-tagger and super-taggers, and then only to introduce weights connecting a new lexical item with its corresponding tag. Both taggers make use of many additional features, for example features relating to the dependency of a tag on the two words to either side. However, these additional feature weights do not seem to be effectively estimated by the approach we consider here. Instead, we focus on estimating the feature weights that correspond to the likelihood of a given word taking a particular tag.

3.3 EasyCCG

EasyCCG[7] (Lewis and Steedman, 2014) is another CCG-based parser that also relies on a log-linear model, as described by Equation 4, but only within what is essentially its super-tagger. POS-tagging is avoided as it represents a bottle-neck within the C&C parser, with wrongly assigned POS tags being difficult to recover from. Similarly, the probabilistic model of parse trees is discarded, and instead an A* parser (Klein and Manning, 2003a) is used to search for the valid CCG derivation that maximises the probabilities of the categories assigned to words in the input. The effectiveness of

[6] http://svn.ask.it.usyd.edu.au/trac/candc

[7] http://homepages.inf.ed.ac.uk/s1049478/easyccg.html

this approach depends both on the constraints imposed on derivations by the CCG formalism and also on the performance of the super-tagger, with the latter aspect being reliant on the features chosen for this model.

Whereas the features used by the C&C parser are structures that are explicitly present in the training data, such as a particular sequence of tags or a CCG rule that involves particular head and dependent words, EasyCCG uses low-dimensional word vectors as features, alongside more traditional features such as capitalisation and 2-character suffixes. The CCG category of an input token is then predicted by a log-linear classifier using the features in a 7-word window surrounding it. The word vectors are initialised using the 50-dimensional embeddings induced by Turian (2010) on 37 million words of newswire text, and are further optimised during training on CCGbank. The use of these word vectors allows EasyCCG to generalise well to out-of-domain data, both because embeddings are available for a wider vocabulary than is found in CCGbank and also because the low dimensionality of the vectors counters some of the problems of sparsity.

4 Evaluation

We measure the performance of our parsers in terms of the ability to recover dependencies from biomedical text. Dependency recovery is not only a useful component in processing both clinical text (Lewis et al., 2011; Sohn et al., 2012) and biomedical literature (Seoud and Mabrouk, 2013; Cohen and Elhadad, 2012; Miyao et al., 2008; Poon and Vanderwende, 2010; Qian and Zhou, 2012), it also provides an evaluation metric that is independent of the particular syntactic formalism employed in the parser.

BioInfer (Pyysalo et al., 2007b) is a corpus of about 35,000 words from PUBMED abstracts, annotated with grammatical relations using a slight modification of the Stanford dependencies scheme (de Marneffe et al., 2006). Our models were tuned on a development set of 600 sentences and then evaluated on the remaining 500 sentence test set, using the same split as Pyysalo et al. (2007a) and Rimmel and Clark (2009). The vocabulary in these sentences diverges considerably from that found in the WSJ, with about 27% of the tokens being unseen. Of the $\approx 3,000$ unseen word types found in BioInfer, 92% occur in the unlabelled Medline corpus that we use to induce distributional representations, and over 80% are assigned parameters by the KNN method. In contrast, only about 700 of those unseen words are present in the SENNA vocabulary, all of which are assigned parameters.

Parser	Type	D	k	F-Score
Berkeley	original	-	-	70.67
	2gram	**200**	**10**	**71.37**
	3gram	50	10	70.55
	4gram	2000	5	69.76
	BOW	50	5	70.12
	SG-bio	300	5	68.44
	SENNA	50	10	70.49
	SG-news	300	16	70.41
C&C	original	-	-	76.39
	2gram	200	4	77.52
	3gram	**500**	**3**	**77.82**
	4gram	2000	3	77.61
	BOW	50	5	75.95
	SG-bio	300	3	76.26
	SENNA	50	10	77.02
	SG-news	300	1	76.64
EasyCCG	original	-	-	78.23
	2gram	**100**	**7**	**79.16**
	3gram	1000	7	78.78
	4gram	10000	3	79.02
	BOW	10	20	78.11
	SG-bio	300	18	76.80
	SENNA	50	20	78.65
	SG-news	300	10	78.01

Table 1: F-scores for recovery of dependencies on the BioInfer development set for the best performing D and k for each type of KNN model.

5 Results

Table 1 compares the performance of the Berkeley, C&C and EasyCCG parsers on the BioInfer development set, after KNN adaptation using various forms of distributional similarity. The results for each parser are grouped together with the first line in each of these groups giving the baseline F-score achieved on the BioInfer development set before expanding the vocabulary. Each subsequent line then corresponds to the best model found for each type of representation, with columns containing D, the number of dimensions in the distributional vectors, k, the number of nearest neighbours, and lastly the F-Score.

The types of distributional representation used in the KNN algorithm are subdivided into those constructed on our Medline titles and abstracts and those trained by their authors on other data sources before being made publicly available. The former group consist of the ngram contexts (2gram, 3gram and 4gram), the bag-of-words contexts (BOW) and the retrained Skip-gram model (SG-bio). The downloaded Skip-gram (SG-news) and SENNA (SENNA) vectors make up the latter group.

Looking first at the differences between these approaches to constructing distributional representations, it is reasonably clear that within each parser the worst performing models tend to be those based on bag-of-words contexts (BOW, SG-news and SG-bio). Of the neural embedding models, SENNA gets the best performance, which we attribute to its preservation of sequential order in handling context. Surprisingly, the Skip-gram model retrained on biomedical data (SG-bio) fared worse than the original (SG-news), due probably in large part to the fact that the original training data was almost 100 times larger than our 1.2B word corpus. The ngram contexts achieved the best F-Scores fairly consistently for all parsers, vindicating our appeal to the psycholinguistic research of Cartwright and Brent (1997), Mintz (2003) and Redington et al. (1998).

Turning now to each parser individually, the baseline performance of the Berkeley Parser proved difficult to exceed, with only the 2gram distributional contexts giving any improvement. The best model used the 200 most frequent bigrams as contexts and averaged over 10 nearest neighbours to achieve an uplift of only 0.7% in F-Score. All other types of model resulted in the Berkeley Parser's performance degrading. For the C&C parser, in contrast, most types of representation, except SG-bio and BOW, achieved an uplift. The best model used the 500 most frequent 3gram contexts, and 3 nearest neighbours to infer parameters for unseen words, improving the F-Score by 1.43%. In comparison, the EasyCCG models achieve higher F-Scores but show smaller uplifts. Here, the best model is based on 2grams, using only 100 such contexts, but requiring 7 nearest neighbours to raise the F-Score by 0.93%.

The results of applying these best performing models to the BioInfer test set are given in Table 2. We evaluate performance on both the set of all dependencies and also the subset of dependencies involving unseen words only. All parsers show an uplift on both measures, with C&C achieving the greatest gains: 2.13% over the whole test set and 6.44% on unseen words. The other parsers obtain smaller uplifts of around 3% on the unseen words but these OOV improvements are nonetheless significant at $p < 0.01$ on a bootstrap test (Efron and Tibshirani, 1993) for all parsers. The improvements over the whole test set are diluted by comparison, although still positive.

Parser	Model	F-score	
		All	Unseen
Berkeley	original	69.85	52.78
	enhanced	70.17	55.98
C&C	original	75.56	63.84
	enhanced	77.69	70.28
EasyCCG	original	77.19	71.44
	enhanced	78.31	74.15

Table 2: F-scores for recovery of dependencies for the original models and the best performing KNN enhanced models on the BioInfer test set.

6 Discussion

We have demonstrated a KNN algorithm to estimate parameters for new lexical items that produces improvements in F-score of up to 6% in the recovery of dependencies in biomedical text. These improvements were obtained without having to retrain the parsers, based simply on distributional representations constructed on unlabelled corpora. In fact, since the context vectors comprehensively outperformed the neural embeddings, our approach achieved these gains without having to induce a clustering or other model over the unlabelled corpora and required only counts for ngrams containing the seen and unseen words. In principle, this method could be applied on the fly, as and when the parser encounters new vocabulary. The success of this ngram based approach is also consistent with psycholinguistic research into syntactic acquisition (Cartwright and Brent, 1997; Mintz, 2003; Redington et al., 1998)

We were able to assign parameters to over 80% of the unseen word types. This introduction of parameters for new word types into the lexicon was the only modification made to the parsers, with the remainder of the models being left unchanged. When combined with methods that could adapt the existing model parameters to the statistics of the new domain, such as self-training (e.g., Deoskar et al., 2014), we expect further improvements to be achievable.

Nonetheless, there were substantial variations in the strength of the improvement attained, with the weak performance of the Berkeley Parser being a notable disappointment. Several differences could be invoked to explain this shortfall. Firstly, the Berkeley Parser has a strong OOV process, and it may just be difficult to beat the estimates it produces, without seeing gold standard data. Secondly, it is a generative rather than a discriminative model, and this complicates the process of modifying the lexicon with questions of how much probability mass to give to unseen words and how to renormalise the lexicon afterwards. Thirdly, rather than representing a single coherent type of linguistic information, the categories induced by the splitting and merging process are just simply the results of whatever splits happened to give the most improvement during training. An example of a subcategory within DT might differentiate definiteness from indefiniteness, while a subcategory in NNP might separate personal names from place names. The inhomogeneity in the type of information encoded in these subcategories probably contributed to our being unable to find distributional information which could be used to induce useful probabilities for them. Consequently, our KNN parameter induction worked only at the level of POS tags for this parser and was therefore less predictive. Andreas and Klein (2014) also struggled to obtain performance improvements for the Berkeley Parser using a distributional matching method. Their problems were also compounded by using SENNA vectors, which we found to give weaker benefits than the ngram context approach.

Our method has certain aspects in common with other approaches to domain adaptation. For example, Koo et al. (2008) train a dependency parser on features deriving from distributional clusters, with two words having similar cluster features if they have similar bigram distributions. Thus, these clusters engender a form of distributional similarity comparable to that used in our KNN algorithm.

KNN algorithms are also commonly used in Graph-Based Semi-Supervised Learning approaches (Das and Petrov, 2011; Altun et al., 2006; Subramanya et al., 2010), with the k-nearest-neighbour sets determining the edges that structure the graph. POS tags are then propagated through the graph from labelled to unlabelled data. Although similarity in these cases is commonly being assessed between token sequences, as opposed to word types, the features used are similar to the ngram templates used here and the bigram distributions used by Koo et al. (2008).

A major difference in our approach is that it does not require retraining the parser or constructing a full model on the unlabelled data. We simply copy parameters from words in the existing lexicon to unseen words, based on a distributional measure of similarity. Moreover, we don't need to see the entire unlabelled corpus. Instead, we can estimate parameters for an unseen word based simply on a set of ngrams centered on it, along with the corresponding ngrams for the existing lexicon.

A reasonable direction for future work would be to develop the way we select the contexts on which our distributional representations are based. In particular, it would make sense to exploit the approach of Brent (1991) and Manning (1993) in which these contexts have an *a priori* linguistic association with particular syntactic frames, as opposed to a merely empirical association deriving from a k-nearest-neighbour model.

Acknowledgements

We would like to thank our colleagues and reviewers for criticism, advice and discussion. This work was supported by ERC Advanced Fellowship 249520 GRAMPLUS and EU Cognitive Systems project FP7-ICT-270273 Xperience.

References

Yasemin Altun, David McAllester, and Mikhail Belkin. 2006. Maximum margin semi-supervised learning for structured variables. In Y. Weiss, B. Schölkopf, and J. Platt, editors, *Advances in Neural Information Processing Systems 18*, pages 33–40. MIT Press, Cambridge, MA.

Jacob Andreas and Dan Klein. 2014. How much do word embeddings encode about syntax? In *Proceedings of the 52nd Annual Meeting of the Association for Computational Linguistics, ACL 2014, June 22-27, 2014, Baltimore, MD, USA, Volume 2: Short Papers*, pages 822–827. The Association for Computational Linguistics.

Mohit Bansal, Kevin Gimpel, and Karen Livescu. 2014. Tailoring continuous word representations for dependency parsing. In *Proceedings of the 52nd Annual Meeting of the Association for Computational Linguistics, ACL 2014, June 22-27, 2014, Baltimore, MD, USA, Volume 2: Short Papers*, pages 809–815. The Association for Computational Linguistics.

Michael R. Brent. 1991. Automatic acquisition of subcategorization frames from untagged text. In Douglas E. Appelt, editor, *29th Annual Meeting of the Association for Computational Linguistics, 18-21 June 1991, University of California, Berkeley, California, USA, Proceedings*, pages 209–214. Morgan Kaufmann.

Timothy A. Cartwright and Michael R. Brent. 1997. Syntactic categorization in early language acquisition: Formalizing the role of distributional analysis. *Cognition*, 63:121–170.

Eugene Charniak. 2001. Immediate-head parsing for language models. In *Proceedings of the 39th Annual Meeting on Association for Computational Linguistics*, ACL '01, pages 124–131. Association for Computational Linguistics.

Raphael Cohen and Michael Elhadad. 2012. Syntactic dependency parsers for biomedical-NLP. In *AMIA 2012, American Medical Informatics Association Annual Symposium, Chicago, Illinois, USA, November 3-7, 2012*, pages 121–128.

Michael Collins. 2003. Head-driven statistical models for natural language parsing. *Computational Linguistics*, 29(4):589–637, December.

Ronan Collobert, Jason Weston, Léon Bottou, Michael Karlen, Koray Kavukcuoglu, and Pavel Kuksa. 2011. Natural language processing (almost) from scratch. *Journal of Machine Learning Research*, 12:2493–2537.

James Curran, Stephen Clark, and Johan Bos. 2007. Linguistically motivated large-scale NLP with C&C and Boxer. In *Proceedings of the 45th Annual Meeting of the Association for Computational Linguistics Companion Volume Proceedings of the Demo and Poster Sessions*, pages 33–36, Prague, Czech Republic, June. Association for Computational Linguistics.

Dipanjan Das and Slav Petrov. 2011. Unsupervised part-of-speech tagging with bilingual graph-based projections. In Dekang Lin, Yuji Matsumoto, and Rada Mihalcea, editors, *The 49th Annual Meeting of the Association for Computational Linguistics: Human Language Technologies, Proceedings of the Conference, 19-24 June, 2011, Portland, Oregon, USA*, pages 600–609. The Association for Computer Linguistics.

Marie-Catherine de Marneffe, Bill MacCartney, and Christopher D. Manning. 2006. Generating typed dependency parses from phrase structure parses. In *Proceedings of the Fifth International Conference on Language Resources and Evaluation (LREC-2006)*, pages 449–454, Genoa, Italy, May. European Language Resources Association (ELRA). ACL Anthology Identifier: L06-1260.

Tejaswini Deoskar, Christos Christodoulopoulos, Alexandra Birch, and Mark Steedman. 2014. Generalizing a strongly lexicalized parser using unlabeled data. In *In Proceedings of Conference of the European Chapter of the Association for Computational Linguistics*, Gothenburg, Sweden.

B. Efron and R. J. Tibshirani. 1993. *An Introduction to the Bootstrap*. Chapman & Hall, New York, NY.

Julia Hockenmaier and Mark Steedman. 2007. CCGbank: A corpus of CCG derivations and dependency structures extracted from the Penn Treebank. *Comput. Linguist.*, 33(3):355–396, September.

Marc F. Joanisse and Mark S. Seidenberg. 1999. Impairments in verb morphology after brain injury: A connectionist model. *Proceedings of the National Academy of Sciences*, 96:7592.

Dan Klein and Christopher D. Manning. 2003a. A* parsing: Fast exact Viterbi parse selection. In *Proceedings of the 2003 Conference of the North American Chapter of the Association for Computational Linguistics on Human Language Technology - Volume 1*, NAACL '03, pages 40–47. Association for Computational Linguistics.

Dan Klein and Christopher D. Manning. 2003b. Accurate unlexicalized parsing. In Erhard W. Hinrichs and Dan Roth, editors, *Proceedings of the 41st Annual Meeting of the Association for Computational Linguistics, 7-12 July 2003, Sapporo Convention Center, Sapporo, Japan*, pages 423–430. The Association for Computer Linguistics.

Terry Koo, Xavier Carreras, and Michael Collins. 2008. Simple semi-supervised dependency parsing. In Kathleen McKeown, Johanna D. Moore, Simone Teufel, James Allan, and Sadaoki Furui, editors, *ACL 2008, Proceedings of the 46th Annual Meeting of the Association for Computational Linguistics, June 15-20, 2008, Columbus, Ohio, USA*, pages 595–603. The Association for Computer Linguistics.

Mike Lewis and Mark Steedman. 2014. A* CCG parsing with a supertag-factored model. In *Proceedings of the 2014 Conference on Empirical Methods in Natural Language Processing (EMNLP)*, pages 990–1000, Doha, Qatar, October. Association for Computational Linguistics.

Neal Lewis, Daniel Gruhl, and Hui Yang. 2011. Dependency parsing for extracting family history. In *2011 IEEE International Conference on Healthcare Informatics, Imaging and Systems Biology, HISB 2011, San Jose, CA, USA, July 26-29, 2011*, pages 237–242. IEEE.

Christopher D. Manning. 1993. Automatic acquisition of a large subcategorization dictionary from corpora.

In Lenhart K. Schubert, editor, *31st Annual Meeting of the Association for Computational Linguistics, 22-26 June 1993, Ohio State University, Columbus, Ohio, USA, Proceedings*, pages 235–242. The Association for Computer Linguistics.

Ryan T. McDonald and Fernando C. N. Pereira. 2006. Online learning of approximate dependency parsing algorithms. In *EACL 2006, 11st Conference of the European Chapter of the Association for Computational Linguistics, Proceedings of the Conference, April 3-7, 2006, Trento, Italy*. The Association for Computational Linguistics.

Tomas Mikolov, Ilya Sutskever, Kai Chen, Gregory S. Corrado, and Jeffrey Dean. 2013. Distributed representations of words and phrases and their compositionality. In Christopher J. C. Burges, Léon Bottou, Zoubin Ghahramani, and Kilian Q. Weinberger, editors, *Advances in Neural Information Processing Systems 26: 27th Annual Conference on Neural Information Processing Systems 2013. Proceedings of a meeting held December 5-8, 2013, Lake Tahoe, Nevada, United States.*, pages 3111–3119.

Toben H. Mintz. 2003. Frequent frames as a cue for grammatical categories in child directed speech. *Cognition.*, 90(1):91–117.

Yusuke Miyao, Kenji Sagae, Rune Sætre, Takuya Matsuzaki, and Jun'ichi Tsujii. 2008. Evaluating contributions of natural language parsers to protein-protein interaction extraction. *Bioinformatics*, 25(3):394–400, December.

Slav Petrov and Dan Klein. 2007. Improved inference for unlexicalized parsing. In *Human Language Technology Conference of the North American Chapter of the Association of Computational Linguistics, Proceedings, April 22-27, 2007, Rochester, New York, USA*, pages 404–411. The Association for Computational Linguistics.

Slav Petrov, Leon Barrett, Romain Thibaux, and Dan Klein. 2006. Learning accurate, compact, and interpretable tree annotation. In *ACL 2006, 21st International Conference on Computational Linguistics and 44th Annual Meeting of the Association for Computational Linguistics, Proceedings of the Conference, Sydney, Australia, 17-21 July 2006*. The Association for Computational Linguistics.

Steven Pinker. 1987. The bootstrapping problem in language acquisition. In B. MacWhinney, editor, *Mechanisms of language aquisition*, pages 399–441. Lawrence Erlbaum Assoc, Hillsdale, NJ.

Hoifung Poon and Lucy Vanderwende. 2010. Joint inference for knowledge extraction from biomedical literature. In *Human Language Technologies: The 2010 Annual Conference of the North American Chapter of the Association for Computational Linguistics*, HLT '10, pages 813–821, Los Angeles, CA, USA. Association for Computational Linguistics.

Sampo Pyysalo, Filip Ginter, Katri Haverinen, Juho Heimonen, Tapio Salakoski, and Veronika Laippala. 2007a. On the unification of syntactic annotations under the Stanford dependency scheme: a case study on BioInfer and GENIA. In *Proceedings of the Workshop on BioNLP 2007: Biological, Translational, and Clinical Language Processing*, pages 25–32. Association for Computational Linguistics.

Sampo Pyysalo, Filip Ginter, Juho Heimonen, Jari Bjorne, Jorma Boberg, Jouni Jarvinen, and Tapio Salakoski. 2007b. BioInfer: a corpus for information extraction in the biomedical domain. *BMC Bioinformatics*, 8(1):50–73.

Longhua Qian and Guodong Zhou. 2012. Tree kernel-based protein-protein interaction extraction from biomedical literature. *Journal of Biomedical Informatics*, 45(3):535–543.

Martin Redington, Nick Chater, and Steven Finch. 1998. Distributional information: A powerful cue for acquiring syntactic categories. *Cognitive Science*, 22(4):425–469.

Laura Rimell and Stephen Clark. 2009. Porting a lexicalized-grammar parser to the biomedical domain. *Journal of Biomedical Informatics*, 42(5):852–865.

Rania Abul A. Seoud and Mai S. Mabrouk. 2013. Tmthcc: A tool for text mining the biomedical literature for hepatocellular carcinoma (hcc) biomarkers identification. *Computer methods and programs in biomedicine*, 112(3):640–648, August.

Steven L. Small, John Hart, Tran Nguyen, and Barry Gordon. 1995. Distributed representations of semantic knowledge in the brain. *Brain*, 118.

Sunghwan Sohn, Stephen Wu, and Christopher G Chute. 2012. Dependency parser-based negation detection in clinical narrative. *AMIA Summits on Translational Science Proceedings*, 2012:1–8.

Mark Steedman. 2000. *The Syntactic Process*. MIT Press, Cambridge, MA, USA.

Amarnag Subramanya, Slav Petrov, and Fernando Pereira. 2010. Efficient graph-based semi-supervised learning of structured tagging models. In *Proceedings of the 2010 Conference on Empirical Methods in Natural Language Processing*, pages 167–176. Association for Computational Linguistics.

Yuta Tsuboi. 2014. Neural networks leverage corpus-wide information for part-of-speech tagging. In *Proceedings of the 2014 Conference on Empirical Methods in Natural Language Processing (EMNLP)*, pages 938–950, Doha, Qatar, October. Association for Computational Linguistics.

Joseph Turian, Lev Ratinov, and Yoshua Bengio. 2010. Word representations: A simple and general method for semi-supervised learning. In *Proceedings of the*

48th Annual Meeting of the Association for Computational Linguistics, pages 384–394. Association for Computational Linguistics.

Expanding a dictionary of marker words for uncertainty and negation using distributional semantics

Alyaa Alfalahi[1], Maria Skeppstedt[2,3,*], Rickard Ahlbom[1], Roza Baskalayci[1], Aron Henriksson[1,*], Lars Asker[1], Carita Paradis[4], Andreas Kerren[3]
[1]DSV, Stockholm University, Stockholm, Sweden [2]Gavagai AB, Stockholm, Sweden
[3]Computer Science Department, Linnaeus University, Växjö, Sweden
[4]Centre for Languages and Literature, Lund University, Lund, Sweden
*Corresponding authors: maria@gavagai.se, aronhen@dsv.su.se

Abstract

Approaches to determining the factuality of diagnoses and findings in clinical text tend to rely on dictionaries of marker words for uncertainty and negation. Here, a method for semi-automatically expanding a dictionary of marker words using distributional semantics is presented and evaluated. It is shown that ranking candidates for inclusion according to their proximity to cluster centroids of semantically similar seed words is more successful than ranking them according to proximity to each individual seed word.

1 Introduction

Clinical text, i.e., the narrative sections of health records, has recently received much attention with regards to automatic detection of uncertainty and negation (Uzuner et al., 2011; Velupillai, 2012; Mowery et al., 2014). Methods for automatic detection of which diagnoses and findings are mentioned as negated or uncertain typically rely on a dictionary of marker words, either as a resource for rule-based methods or when constructing features for machine learning (Uzuner et al., 2011). Dictionaries of marker words have previously been constructed by manual annotation or by translation of dictionaries from one language to another (Velupillai et al., 2014). Alternative methods for automating marker word dictionary construction would, however, be useful since manual annotation is time-consuming, and translation results in incomplete dictionaries due to differences between languages in how negation and uncertainty are expressed. The aim of the present study was to explore one such possible method for semi-automatic dictionary expansion: using distributional semantics to extract possible marker words from a large unannotated corpus and, more specifically, attempting to obtain improved performance by applying clustering to the semantic vectors in the resulting semantic space.

Given a dictionary of known uncertainty and negation markers to use as seed words, the task of the system explored here was to rank words not included in the seed dictionary according to their suitability as marker words, with the aim of having good candidates for inclusion in the dictionary among the top-ranked words.

An experiment was carried out to determine if a method whereby words are ranked according to proximity to the centroids of seed word clusters outperforms – in the sense of ranking true marker words higher – a ranking method that instead uses proximity to each individual seed word. The seed words are here represented as vectors comprising word co-occurrence information, created using a model of distributional semantics called random indexing.

2 Background

For the English language, there are a number of large corpora annotated for speculation and negation: bio-medical corpora (Vincze et al., 2008; Uzuner et al., 2011), as well as corpora in other domains (Konstantinova et al., 2012). Systems for detecting negation and speculation are typically constructed by training machine learning models on these corpora (Farkas et al., 2010; Uzuner et al., 2011). For most other languages, there are, however, often only smaller annotated corpora or none at all (Velupillai et al., 2011; Aramaki et al., 2014). In such cases, methods for detecting uncertainty and negation that rely on lexicon/dictionary-matching to lists of marker words for uncertainty or negation are a possible alternative. Such an approach has been shown to perform in line with machine learning methods trained on corpora with fewer training instances (Velupillai et al., 2014; Aramaki et al., 2014).

Proceedings of the Sixth International Workshop on Health Text Mining and Information Analysis (Louhi), pages 90–96,
Lisbon, Portugal, 17 September 2015.

For a dictionary-matching approach, extensive dictionaries of marker words are, however, required, and to build such a resource manually can also be prohibitively expensive. An alternative to creating a dictionary of marker words manually is to use automatic methods for creating lists of candidate words to include in the dictionary. For semi-automatically creating vocabulary resources of other types than marker words, there are a number of previous studies wherein various methods are used. Those that rely on terms being explicitly defined in the text (Hearst, 1992; Yu and Agichtein, 2003; Cohen et al., 2005; McCrae and Collier, 2008; Neelakantan and Collins, 2014) are unlikely to be successful for negation and uncertainty terms. Term extraction methods that measure similarity between words according to how frequently they occur in similar contexts (Lin, 1998), on the other hand, might be more suitable. Such distributional semantic properties are often represented by spatial models, i.e., given a geometric representation in the form of a vector space (Cohen and Widdows, 2009), and there are examples in which such spatial models have been used for vocabulary expansion (Zhang and Elhadad, 2013; Skeppstedt et al., 2013; Henriksson et al., 2014), as well as for related tasks (Jonnalagadda et al., 2012), in the bio-medical domain.

Random indexing is a computationally lightweight method for producing spatial models of distributional semantics (Kanerva et al., 2000; Sahlgren, 2006). Random indexing requires two types of vectors: index vectors, which are used only for semantic space construction, and context vectors, which represent the meaning of words and collectively make up the resulting semantic space. Each unique word w_j in the corpus vocabulary W is assigned an index vector $\vec{w}^i_j$ and a context vector $\vec{w}^c_j$ of dimensionality d. The index vectors are static representations of contexts (here, these are unique words) that are approximately uncorrelated to each other, which is achieved by creating very sparse vectors that are randomly assigned a small number of non-zero elements (1s and -1s). A $\vec{w}^c_j$ – containing the distributional profile of the word w_j – is then the (weighted) sum of all the index vectors of the words with which w_j co-occurs within a (typically symmetric) window of a certain size. Spatial proximity between two context vectors is taken to indicate the semantic similarity between the two words they represent. The context vectors can also be further analysed, for instance by applying different kinds of clustering (Rosell et al., 2009; Pyysalo et al., 2013).

3 Method

The conducted experiment consisted of the following steps: 1) constructing a semantic space with random indexing; 2) applying hierarchical clustering to context vectors representing seed words; 3) for different levels in the cluster tree, producing a ranked list of the words in the corpus according to their proximity to the centroids of the constructed clusters; 4) evaluating the recall of the top-ranked words in the produced lists against a reference standard.

1) A semantic space was constructed with random indexing on a freely available subset (years 1996–2005) of the *Läkartidningen* (Journal of the Swedish Medical Association) corpus (Kokkinakis, 2012). This subset contains 21,447,900 tokens and 444,601 unique terms. In order also to allow inflected forms of marker words to be captured, the corpus was not lemmatised. 1,000-dimensional vectors were used in a context window of two preceding and two following words and double weight was given to the two words closest to the target word. Since the sentences in the corpus appear in a randomised order, no context windows were allowed to cross sentence boundaries.

2) Single-linkage agglomerative hierarchical clustering (Sibson, 1973) was applied to the context vectors representing the seed words. A tree-formed cluster hierarchy was thereby created, with progressively larger clusters, starting from clusters in which each seed word formed its own cluster (cluster level *0* on the x-axis in Figure 1), until all seed words collectively formed a single cluster (cluster level *79* on the x-axis in Figure 1).

3) For each cluster level (0 to 79), a ranked list of all words in the corpus (except those used as seed words) was produced. The words were ranked according to the Euclidean distance between their length-normalised context vector and their most closely located cluster centroid (also length-normalised). That is, the word with the context vector that was closest to any of the centroid vectors achieved the highest ranking, the word with the context vector that was second closest to any of the centroid vectors was ranked as number two on the list, and so on. For cluster level

Cluster level 0	Cluster level 40	Cluster level 79
misstänka (suspect)	risken (the risk)	barnet (the child)
sannolikt (likely)	analys (analysis)	folk (people)
angeläget (pressing)	påvisats (proven)	arbetsgivaren (the employer)
rimligt (reasonable)	acceptera (accept)	så (so)
förmodligen (probably)	riskerar (risks)	uppdraget (the assignment)
tycker (think)	registrering (registration)	personalen (the staff)
kontrollera (check)	använda (use)	verksamhetscheferna (the business managers)
hävda (assert)	läran (doctrine)	medlet (the agent)
kartlägga (survey)	kommer (come)	läkarna (the doctors)
värdera (estimate)	kunskapen (knowledge)	landstingen (the counties)

Table 1: Top 10 words retrieved for a randomly selected seed word sampling (among the 500 re-samplings used in the experiment. The top 10 words for cluster level 0, 40 and 79 are shown).

0, in which each seed word formed its own cluster, the centroids were composed of the context vectors for the seed words, and the words were thus ranked according to their proximity to any of the seed words.

4) As a final step, the method was evaluated using an existing, freely available, dictionary of Swedish marker words for uncertainty and negation. This dictionary was developed through translation of English marker words and through manual annotation of clinical text (Velupillai et al., 2014). Markers in the dictionary were used as seed words as well as for evaluation data.

The dictionary was filtered by removing multi-word terms, since the constructed semantic space only contains single-word terms. In addition, words occurring fewer than 50 times in the corpus were removed, since a certain number of observations of a word is required for its context vector to be modeled reliably in semantic space. The performed filtering resulted in a set of 161 marker words for uncertainty and negation. The vocabulary used is shown in Figure 3.

This set of vocabulary terms was used in the evaluation by randomly splitting it into two equally large subsets: one set of seed words and one set of words to use as reference standard. The set of seed words represents words that, in a real-world scenario, would be included in an existing, but incomplete, dictionary of marker words, and the reference standard represents words that should be included as top-ranked candidates by the evaluated system. The performance of the system was evaluated through a standard information retrieval measure, i.e., by calculating recall (for the n top-ranked candidates) of the produced list against the words in the reference standard. Recall was calculated for up to top 5,000 candidate words (from top 100 with a step size of 100). Candidate list precision for the automatic evaluation is not reported, as this is separated only by a constant from recall, and would therefore show the same pattern with respect to cluster sizes.

To make the results less dependent on which terms were used as seed words and which were used as reference standard words, the experiment was repeated 500 times, each time with a new random split of the 161 words in the dictionary into a seed words set and reference standard set. The final results were achieved by averaging the achieved recall results.

Table 1 shows an example of the top 10 candidates retrieved for one randomly selected seed sample among the 500 evaluated re-samplings. In this short list, and for this sample, there are better candidates for cluster level 0 than for the other cluster levels.

4 Results and Discussion

As can be seen in Figure 1, results achieved with a moderate cluster level (20–40) were better than those achieved when proximity to each individual seed word was used as the ranking method (level 0). When the clusters grew larger (cluster level > 50), however, recall started to decrease, and using proximity to the centroid of a cluster containing all seed words resulted in much lower recall than when using proximity to each individual seed word, indicating that there are important differences in the usage of marker words. As a method for ranking the words in the corpus, it was thus better to use proximity to the centroid of a

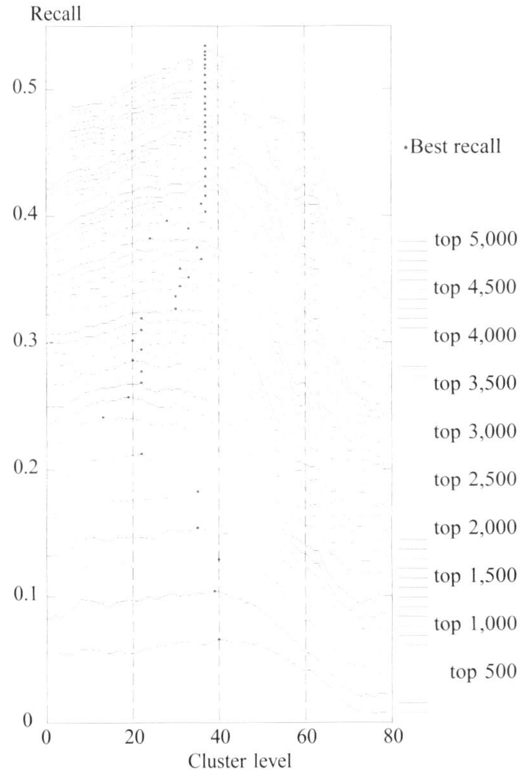

Figure 1: Recall for different levels of clustering. Cluster level 0 means that each seed word forms its own cluster. The higher the cluster level, the larger the clusters created. Cluster level 79 means that all seed words form one large cluster.

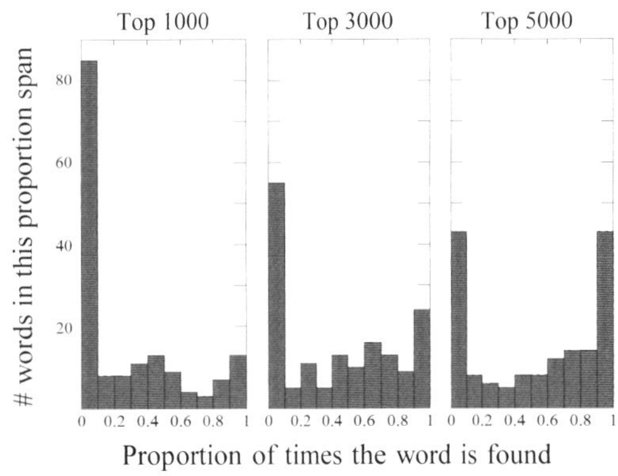

Figure 2: Histogram over the proportion of times a word is found when used as a reference standard word. The first stack shows the number of words that are found between 0% and 10% of the times they are used in the reference standard. The second stack shows the number of words found between 10% and 20% of the times, and so on. The statistics are shown for top 1,000, 3,000 and 5,000 candidates (using the cluster level optimal for top 3,000).

number of semantically similar words than to use proximity to each individual word. When using large clusters of seed words, however, distributionally dissimilar words, e.g., *förnekar (denies)* and *möjlig (possible)*, were clustered together, which decreased recall.

Recall is shown in Figure 1 from among the top 100 best candidates up to among the top 5,000 best candidates (with a step size of 100). The improvement that is achieved with a larger number of candidate words slowly levels out with an increasing number of candidates. The average result among the top 5,000 best candidates was a recall of just above 50%. A possible reason for these relatively low recall scores could be that the dictionary of marker words for uncertainty and negation contains many semantic outliers, i.e., words that do not occur in contexts similar to the other words in the list. The statistics shown in Figure 2 support this theory. The first stack in each of the three histograms, which shows the number of words that are very rarely found, is large in all three histograms. This indicates that regardless of which seed words are used, there is a large number of words that are never or very rarely found. It might, therefore, be the case that methods based on distributional semantics cannot be used for constructing a complete dictionary of negation and uncertainty markers, as such a dictionary includes semantic outliers, although the methods are useful for expanding a dictionary with typical marker words. Figure 3 shows the vocabulary used and how often a word was retrieved among the top 1,000 candidates when used as evaluation data.

It should be noted that the used list of marker words has been constructed using clinical text and has the aim of being used for clinical text, while this study was carried out on medical journal text. The used medical corpus has the advantage of being freely available, in contrast to large clinical corpora, which are only rarely available for research, and it also makes it possible for anyone to repeat the experiments carried out in this study. As there are many differences between medical journal text and clinical text (Smith et al., 2014), some marker words might be used in other contexts in clinical text than in medical journal text,

övertygande(convincing):0.0 överväga(consider):0.82 övervägas(considered):0.0 aldrig(never):0.0 alternativ(option):0.0 alternativa(alternative):0.0
alternativt(alternatively):0.43 angående(relating):0.0 anse(deem):0.97 ansetts(considered):0.0 antagits(been guessed):0.0 antas(is-guessed):0.21 antingen(either):0.0
antogs(was guessed):0.0 antydan(hint):0.0 antyder(implies):0.98 antytt(hinted):0.55 avfärda(dismiss):0.0 avfärdar(dismisses):0.0 beaktande(regard):0.0
bedömning(assessment):0.47 betänka(reports):0.48 borta(gone):0.0 differentialdiagnos(differential-diagnosis):0.0 ej(not):0.0 eventuell(possible):0.3 eventuella(any):0.0
eventuellt(optionally):0.0 förefaller(appears):0.4 föreslå(propose):0.92 föreslår(proposes):0.16 föreslagit(proposed):0.55 förmoda(surmise):0.8 förmodad(putative):0.0
förmodade(putative):0.0 förmodas(believed):0.28 förmodligen(probably):0.76 förneka(deny):0.97 förnekar(denies):0.08 förslagsvis(tentatively):0.4 fråga(issue):0.0
frågan(the-issue):0.06 frågeställning(issue):0.0 frågeställningen(the-issue):0.0 framstår(stands):0.66 framträder(stands):0.0 fri(free):0.0 fria(free):0.0 funderingar(speculations):0.0
granskning(review):0.33 indicerat(indicated):0.0 indikation(indication):0.31 indikationen(the-indication):0.59 indikationer(indications):0.27
indikationerna(the-indications):0.61 indikera(indicate):0.0 indikerar(indicates):0.97 indikerat(indicated):0.43 inga(no):0.0 ingen(no):0.22 ingenting(nothing):0.02
inget(no):0.21 inte(not):0.0 känna(feel):0.0 kanske(maybe):0.74 löst(solved):0.0 liknade(similar):0.02 liknar(resembles):0.3 märka(notice):0.96
möjlig(possible):0.13 möjliga(possible):0.0 möjligen(possibly):0.14 möjligheten(possibility):0.91 möjligt(possible):0.02 möjligtvis(possibly):0.14
misstänka(suspect):0.9 misstänker(suspect):0.57 misstänkt(suspect):0.0 misstänkta(suspects):0.0 misstankar(suspicions):0.64 misstanke(suspicion):0.36
misstanken(suspicion):0.58 negativ(negative):0.13 negativa(negative):0.0 negativt(negative):0.0 nog(probably):0.19 observerades(observed):0.0 observerats(observed):0.0
och/eller(and/or):0.0 oklar(unclear):0.53 oklart(unclear):0.45 oroande(worrying):0.05 osäker(unsure):0.52 osäkerhet(uncertainty):0.0 osäkert(uncertain):0.35
osannolik(improbable):0.0 osannolikt(improbable):0.51 otroligt(incredible):0.0 otydliga(unclear):0.02 påstår(states):1.0 preliminär(provisional):0.0
preliminärt(preliminary):0.0 protokoll(protocol):0.0 protokollet(protocol):0.0 representerar(represents):0.0 rimligtvis(reasonably):0.0 saknar(lack):0.0 saknas(missing):0.0
sannolik(probable):0.47 sannolika(probable):0.4 sannolikheten(probability):0.18 sannolikt(likely):0.81 sett(seen):0.0 stödjer(supports):0.03 svårbedömd(hard-to-assess):0.42
svårtolkade(difficult-to-interpret):0.02 syns(visible):0.0 tendens(tendency):0.0 tendenser(trends):0.0 tolka(interprete):0.98 tolkades(was-interpreted):0.59
tolkar(interpretes):0.0 tolkas(interpreted):0.0 tolkats(interpreted):0.05 torde(should):0.36 tro(believe):0.91 trodde(thought):0.83 trolig(probable):0.34
troliga(probable):0.22 troligen(probably):0.81 troligt(likely):0.22 troligtvis(probably):0.71 tror(think):0.03 tros(believed):0.0 trott(imagined):0.0 tveksam(passable):0.0
tveksamhet(hesitancy):0.0 tveksamt(doubtful):0.14 tycker(think):0.06 tycks(appears):0.46 tydligen(apparently):0.36 undersökning(study):0.37 uppenbarligen(obviously):0.41
uppleva(experience):0.9 upplevd(perceived):0.0 upplevdes(perceived):0.64 upplever(experiencing):0.0 utan(without):0.0 uteslöt(excluded):0.0
utesluta(exclude):0.87 uteslutas(excluded):0.0 utesluter(excludes):0.0 uteslutet(precluded):0.01 uteslutit(excluded):0.0 uteslutits(excluded):0.0
utreda(investigate):0.91 utredning(investigation):0.47 utvärdering(evaluation):0.47 varken(neither):0.0 verkar(seems):0.3 visa(show):0.94

Figure 3: The vocabulary used for the experiments, displayed in a font size corresponding to how often a word, when included in the evaluation data, was retrieved among the top 1,000 candidates. Words displayed in black were retrieved in less than 10% of the times they were included in the evaluation data.

and there might be fewer semantic outliers if the experiments were to be repeated using a clinical corpus.

There were also 54 negation and uncertainty markers in the used dictionary that were excluded from the study since they occurred fewer than 50 times in the corpus. The existence of these words, which were mainly inflected forms, abbreviations and a few misspellings that are unusual outside of the clinical language, e.g., *beaktandes (taking into consideration)*, *alt (alternatively)*, *diffdiagnos (differential diagnosis)*, is also a reason for why the experiment should be repeated with a clinical corpus. Multi-word terms formed an even larger proportion of the terms excluded from the negation and uncertainty dictionary when constructing the vocabulary used in the experiments (376 terms). There are previous studies in which multi-word negation and uncertainty markers have been constructed from single-word markers (Velupillai et al., 2014), but an alternative could be to directly model multi-word terms in semantic space (Henriksson et al., 2013a; Henriksson et al., 2013b).

A manual evaluation of a Swedish uncertainty and negation marker candidate list, produced with the methods of this study, could also be carried out in order to determine to what extent it is possible to obtain words not yet included in the dictionary using this method. The dictionary used for evaluation was, however, obtained by translation of English marker words and by extracting markers from clinical text in which 2,500 diagnostic statements had been annotated (Velupillai et al., 2014). It could, therefore, be difficult to retrieve standard language single-word terms for negation and uncertainty not already included in this dictionary. There might, however, still be a need to add abbreviated forms and multi-word terms. The methods evaluated here could also be applied to other languages, for which resources of marker words for negation and uncertainty, used in medical text, have not yet been constructed.

5 Conclusion

It was shown that proximity to the centroid of a number of semantically similar seed words was a more successful method for ranking the words in the corpus as candidates for negation and uncertainty markers than to use proximity to each individual seed word as the ranking method. However, many of the marked words used in the evaluation were never, or very rarely, ranked highly on the candidate list, regardless of which seed words were used.

Acknowledgements

This work was partly funded through the project StaViCTA by the framework grant "the Digitized Society Past, Present, and Future" with No. 2012-5659 from the Swedish Research Council (Vetenskapsrådet) and partly by the Swedish Foundation for Strategic Research through the project High-Performance Data Mining for Drug Effect Detection (ref. no. IIS11-0053) at Stockholm University, Sweden. The authors would also like to direct thanks to the reviewers for valuable comments.

References

Eiji Aramaki, Mizuki Morita, Yoshinobu Kano, and Tomoko Ohkuma. 2014. Overview of the NTCIR-11 MedNLP-2 task. In *Proceedings of NTCIR-11*.

Trevor Cohen and Dominic Widdows. 2009. Empirical distributional semantics: Methods and biomedical applications. *Journal of Biomedical Informatics*, 42(2):390 – 405.

Aaron Cohen, William R. Hersh, Christopher Dubay, and Kent Spackman. 2005. Using co-occurrence network structure to extract synonymous gene and protein names from medline abstracts. *BMC Bioinformatics*, 6(1):103.

Richárd Farkas, Veronika Vincze, György Móra, János Csirik, and György Szarvas. 2010. The CoNLL-2010 shared task: Learning to detect hedges and their scope in natural language text. In *Proceedings of the Fourteenth Conference on Computational Natural Language Learning*, pages 1–12.

Marti Hearst. 1992. Automatic acquisition of hyponyms from large text corpora. In *Proceedings of COLING 1992*, pages 539–545.

Aron Henriksson, Mike Conway, Martin Duneld, and Wendy W. Chapman. 2013a. Identifying synonymy between SNOMED clinical terms of varying length using distributional analysis of electronic health records. In *Proceedings of the Annual Symposium of the American Medical Informatics Association (AMIA 2013)*, Washington DC, USA.

Aron Henriksson, Maria Skeppstedt, Maria Kvist, Martin Duneld, and Mike Conway. 2013b. Corpus-Driven Terminology Development: Populating Swedish SNOMED CT with Synonyms Extracted from Electronic Health Records. In *Proceedings of the 2013 Workshop on Biomedical Natural Language Processing*, pages 36–44, Sofia, Bulgaria. Association for Computational Linguistics.

Aron Henriksson, Hans Moen, Maria Skeppstedt, Vidas Daudaravičius, and Martin Duneld. 2014. Synonym extraction and abbreviation expansion with ensembles of semantic spaces. *J Biomed Semantics*, 5(1):6.

Siddhartha Jonnalagadda, Trevor Cohen, Stephen Wu, and Graciela Gonzalez. 2012. Enhancing clinical concept extraction with distributional semantics. *J Biomed Inform*, 45(1):129–40, Feb.

Pentti Kanerva, Jan Kristoferson, and Anders Holst. 2000. Random indexing of text samples for latent semantic analysis. In L. R. Gleitman and A. K. Joshi, editors, *Proceedings of the 22nd Annual Conference of the Cognitive Science Society*, Mahwah, NJ.

Dimitrios Kokkinakis. 2012. The journal of the Swedish medical association - a corpus resource for biomedical text mining in Swedish. In *The Third Workshop on Building and Evaluating Resources for Biomedical Text Mining (BioTxtM), an LREC Workshop. Turkey.*

Natalia Konstantinova, Sheila C.M. de Sousa, Noa P. Cruz, Manuel J. Maña, Maite Taboada, and Ruslan Mitkov. 2012. A review corpus annotated for negation, speculation and their scope. In Nicoletta Calzolari, Khalid Choukri, Thierry Declerck, Mehmet Uğur Doğanur, Bente Maegaard, Joseph Mariani, Jan Odijk, and Stelios Piperidis, editors, *Proceedings of the Eight International Conference on Language Resources and Evaluation (LREC'12)*, Istanbul, Turkey. European Language Resources Association (ELRA).

Dekang Lin. 1998. Automatic retrieval and clustering of similar words. In *Proceedings of the 36th Annual Meeting of the Association for Computational Linguistics and 17th International Conference on Computational Linguistics–Volume 2*, ACL '98, pages 768–774, Stroudsburg, PA, USA. Association for Computational Linguistics.

John McCrae and Nigel Collier. 2008. Synonym set extraction from the biomedical literature by lexical pattern discovery. *BMC Bioinformatics*, 9:159.

Danielle Mowery, Sumithra Velupillai, Brett R. South, Lee Christensen, David Martinez, Liadh Kelly, Lorraine Goeuriot, Noemie Elhadad, Sameer Pradhan, Guergana Savova, and Chapman Wendy W. 2014. Task 2: ShARe/CLEF eHealth Evaluation Lab 2014. In L. Cappellato, N. Ferro, M. Halvey, and W. Kraaij, editors, *CLEF2014 Working Notes*, volume 1180, pages 31–42. CEUR-WS, September.

Arvind Neelakantan and Michael Collins. 2014. Learning dictionaries for named entity recognition using minimal supervision. In Gosse Bouma and Yannick Parmentier, editors, *Proceedings of the 14th Conference of the European Chapter of the Association for Computational Linguistics, EACL 2014, April 26-30, 2014, Gothenburg, Sweden*, pages 452–461. The Association for Computer Linguistics.

Sampo Pyysalo, Filip Ginter, Hans Moen, Tapio Salakoski, and Sophia Ananiadou. 2013. Distributional semantics resources for biomedical text processing. In *Proceedings of Languages in Biology and Medicine.*

Magnus Rosell, Martin Hassel, and Viggo Kann. 2009. Global evaluation of random indexing through Swedish word clustering compared to the people's dictionary of synonyms. In *Proceedings of Recent Advances in Natural Language Processing*, Borovets, Bulgaria.

Magnus Sahlgren. 2006. *The Word-Space Model: Using distributional analysis to represent syntagmatic and paradigmatic relations between words in high-dimensional vector spaces.* Doctoral thesis, Stockholm University.

R. Sibson. 1973. SLINK: an optimally efficient algorithm for the single-link cluster method. *Comput. J.*, 16(1):30–34.

Maria Skeppstedt, Magnus Ahltorp, and Aron Henriksson. 2013. Vocabulary expansion by semantic extraction of medical terms. In *Proceedings of Languages in Biology and Medicine (LBM)*, Tokyo, Japan, December.

Kelly Smith, Beata Megyesi, Sumithra Velupillai, and Maria Kvist. 2014. Professional language in Swedish clinical text: Linguistic characterization and comparative studies. *Nordic Journal of Linguistics*, 37:297–323.

Özlem. Uzuner, Brett R. South, Shuying Shen, and Scott L. DuVall. 2011. 2010 i2b2/va challenge on concepts, assertions, and relations in clinical text. *J Am Med Inform Assoc*, 18(5):552–556.

Sumithra Velupillai, Hercules Dalianis, and Maria Kvist. 2011. Factuality Levels of Diagnoses in Swedish Clinical Text. In A. Moen, S. K. Andersen, J. Aarts, and P. Hurlen, editors, *Proc. XXIII International Conference of the European Federation for Medical Informatics (User Centred Networked Health Care)*, pages 559–563, Oslo, August. IOS Press.

Sumithra Velupillai, Maria Skeppstedt, Maria Kvist, Danielle Mowery, Brian E Chapman, Hercules Dalianis, and Wendy W Chapman. 2014. Cue-based assertion classification for swedish clinical text–developing a lexicon for pycontextswe. *Artif Intell Med*, 61(3):137–44, Jul.

Sumithra Velupillai. 2012. *Shades of Certainty – Annotation and Classification of Swedish Medical Records*. Doctoral thesis, Department of Computer and Systems Sciences, Stockholm University, Stockholm, Sweden, April.

Veronika Vincze, György Szarvas, Richárd Farkas, György Móra1, and János Csirik. 2008. The BioScope Corpus: Biomedical texts annotated for uncertainty, negation and their scopes. *BMC Bioinformatics*, 9(Suppl 11):S9.

Hong Yu and Eugene Agichtein. 2003. Extracting synonymous gene and protein terms from biological literature. *Bioinformatics*, 1(19):340–349.

Shaodian Zhang and Noémie Elhadad. 2013. Unsupervised biomedical named entity recognition: Experiments with clinical and biological texts. *Journal of Biomedical Informatics*, 46(6):1088–1098. Special Section: Social Media Environments.

Held-out versus Gold Standard: Comparison of Evaluation Strategies for Distantly Supervised Relation Extraction from Medline abstracts

Roland Roller and **Mark Stevenson**
Department of Computer Science
University of Sheffield
Regent Court, 211 Portobello
S1 4DP Sheffield, England
roland.roller,mark.stevenson@sheffield.ac.uk

Abstract

Distant supervision is a useful technique for creating relation classifiers in the absence of labelled data. The approaches are often evaluated using a held-out portion of the distantly labelled data, thereby avoiding the need for lablelled data entirely. However, held-out evaluation means that systems are tested against noisy data, making it difficult to determine their true accuracy. This paper examines the effectiveness of using held-out data to evaluate relation extraction systems by comparing the results that are produced with those generated using manually labelled versions of the same data. We train classifiers to detect two UMLS Metathesaurus relations (*may-treat* and *may-prevent*) in Medline abstracts. A new evaluation data set for these relations is made available. We show that evaluation against a distantly labelled gold standard tends to overestimate performance and that no direct connection can be found between improved performance against distantly and manually labelled gold standards.

1 Introduction

Relation extraction is a popular topic in the biomedical domain and has been the subject of several challenges (e.g. DDI challenge (Segura-Bedmar et al., 2013), BioNLP Shared Task (Nédellec et al., 2013)). Many approaches rely on supervised learning techniques using manually labelled training data. However, the creation of annotated training data is time-consuming, expensive and often requires expert knowledge.

Distant supervision (self-supervised learning) is a widely applied technique for training relation extraction systems (Wu and Weld, 2007; Krause et al., 2012; Roth and Klakow, 2013; Ritter et al., 2013; Vlachos and Clark, 2014) that avoids the need for annotated training data. Training examples are annotated automatically using a knowledge base. Facts from the knowledge base are matched against text and used as training examples. For example, a knowledge base may assert that the entity pair *CONDITION("hair loss")-DRUG("paroxetine")* is an instance of the relationship *adverse-drug effect*. Distant supervision approaches normally assume that sentences containing both entities assert the relation between them and, consequently, the following sentence would be used as a positive example of the *adverse-drug effect* relation:

> *"Findings on discontinuation and rechallenge supported the assumption that the* ***hair loss*** *was a side effect of the* ***paroxetine****." (PMID=10442258)*

However, this assumption does not always hold which can lead to sentences containing entity pairs being mistakenly identified as asserting a particular relation between them. For example, the following sentence contains the same entity pair but does not assert the *adverse-drug effect* relation:

> *"There are a few case reports on* ***hair loss*** *associated with tricyclic antidepressants and serotonin selective reuptake inhibitors (SSRIs), but none deal specifically with* ***paroxetine****." (PMID=10442258)*

Consequently, data annotated using distant supervision is noisy and unlikely to be of as high

Proceedings of the Sixth International Workshop on Health Text Mining and Information Analysis (Louhi), pages 97–102,
Lisbon, Portugal, 17 September 2015.

quality as manually labelled data. Despite this distantly supervised relation extraction provides reasonable results compared to those based on supervised learning (see e.g. in (Thomas et al., 2011)).

Distant supervision allows relation extraction systems to be created without manually labelled data. However, this raises the issue of how such a system can be evaluated. Previous approaches have carried out evaluation using existing data sets labelled with examples of the target relation (Bellare and Mccallum, 2007; Nguyen and Moschitti, 2011; Min et al., 2013) or a similar relation (Thomas et al., 2011; Roller and Stevenson, 2014). However, in the majority of scenarios the best use for any labeled data available is as training data. Others, such as Craven and Kumlien (1999), generated their own gold standard to annotate relevant relations of their knowledge base. But the effort required to generate manually labelled evaluation data somewhat negates the benefit of reduced development time provided by distant supervision.

An alternative approach, which does not require any labelled data, is held-out evaluation. This approach splits facts from the knowledge base into two parts: one to generate distantly supervised training data and the other to generate distantly supervised evaluation data (Mintz et al., 2009; Riedel et al., 2010; Hoffmann et al., 2010; Roller et al., 2015).

This approach is often combined with a manual evaluation in which a subset of the predictions is selected to be examined in more detail. For example, Riedel et al. (2010) supplemented the held-out evaluation of their distant supervision approach for Freebase by selecting the top 1000 facts it predicted and evaluating them manually. Others such as Surdeanu et al. (2012) and Intxaurrondo et al. (2013) work with the same knowledge base and are able to re-use the manually labelled data generated by Riedel et al. (2010). However, this data is only available for some Freebase relations and evaluation data has to be generated for each new relation. Approaches such as Takamatsu et al. (2012), Zhang et al. (2013) and Augenstein et al. (2014) combine a held-out evaluation with a manual evaluation of a randomly chosen subset or the top-k predictions. This technique is a more reliable evaluation method but requires more effort including (potentially) domain knowledge and needs to be repeated for each version of the classifier.

Held-out evaluation using distantly labelled data is a simple and quick technique for estimating the accuracy of distantly supervised relation extraction systems. However, this evaluation data is noisy and it is unclear what effect this has on the accuracy of performance estimates.

The issue is explored in this paper by evaluating relation extraction systems for two biomedical relations using both manually and distantly labelled data. We automatically generate labelled held-out data and then carry out a manual annotation to allow direct comparison. A distantly supervised classifier is trained and evaluated on both data sets. Similar as in Xu et al. (2013) we show that a large portion of the labels generated by distant supervision for the two relations are incorrect. However we find that evaluating classifiers using held-out distantly supervised data tends to overestimate performance compared to manually labelled data and that improvements in performance observed in evaluation against distantly supervised data are not necessarily reflected in improved results when measured against manually labelled data. To the best of our knowledge this is the first direct comparison of evaluating distantly supervised classifiers against distantly and manually labelled gold standards. Analysis in previous work has been restricted to determining the true labels for a set of positively predicted labels.

The remainder of this paper is structured as follows. The next section 2 describes the creation of the distantly supervised data and a manually labelled subset. A comparison of the automatically and manually generated labels is carried out in Section 3. Sections 4 evaluates a relation extraction system using different data sets and compares the performance obtained. The paper concludes with section 5.

2 Data Generation

A large set of distantly labelled examples was generated (Section 2.1). A small portion of these were used as held-out test data. This data set was also manually annotated (Section 2.2).

2.1 Distant labelling

Distantly labelled examples are generated using the Unified Medical Language System (UMLS) Metathesaurus as a knowledge source. UMLS is a large biomedical knowledge base which contains information about millions of medical concepts and the relations between them, making it well

		distantly labelled (DL)					
		may-treat			may-prevent		
		pos	neg	#	pos	neg	#
manually labelled (ML)	pos	**106**	67	173	**85**	54	139
	neg	94	**133**	227	115	**146**	261
		200	200		200	200	

Table 1: Comparison of manual and distantly labelled annotations

suited for distant supervision. Two biomedical relations (*may-treat* and *may-prevent*) were selected from UMLS. These relations describe connections between a pharmacological substance (e.g. drug) and a disease. For example, the following sentence expresses a *may-prevent* relationship between the entities *fluoride* and *dental caries*:

> *"Although **fluoride** is clearly a major reason for the decline in the prevalence of **dental caries**, there are no studies of the incremental benefit of in-office fluoride treatments for low-risk patients exposed to fluoridated water and using fluoridated toothpaste." (PMID=10698247)*

Training data for the two relations was generated from approximately 1 million biomedical abstracts from Medline[1] annotated with UMLS concepts by MetaMap[2] (Aronson and Lang, 2010). Sentences containing concepts that are identified as being related in the UMLS's `MRREL` table were selected and used as positive examples.[3] Negative examples were generated using a closed word assumption: pairs of concepts that are not listed as being related in UMLS for a given relation are considered to be negative examples of that relation. Such pairs are generated by considering all possible pairs from a particular relation and creating new pairs from the set of entities.

2.2 Test Data

A set of 400 distantly labelled sentences were randomly selected for each relation to generate held-out test data. Although the distantly labelled data contains more negatively labelled sentences than positive ones, equal numbers of positive and negative examples (200 of each) are selected in order to ensure that a sufficient number of positive instances are included in the data set. The sentences in this data set were selected so that none of the instance pairs occur in the data used for training. We refer to this data set as **DL** (Distantly Labelled).

The DL data set was then manually annotated. Two annotators were recruited, both of whom were studying graduate degrees in subjects related to medicine at our institution. Given a sentence with a highlighted pharmacological substance and a highlighted disease, the annotators had to determine whether a sentence expresses the relationship of interest between two presented entities or not. The annotators were not shown the labels generated by the distant supervision process. The annotators were asked to only label sentences as positive if it contains a clear indication that the pharmacological substance either treats or prevents the disease. For example, the following sentence mentions that a study has been carried out to determine whether the drug *voriconazole* treats *paracoccidioidomycosis*:

> *"A pilot study was conducted to investigate the efficacy, safety, and tolerability of **voriconazole** for the long-term treatment of acute or chronic **paracoccidioidomycosis**, with itraconazole as the control treatment." (PMID=17990229)*

However, the sentence does not contain any indication that the drug successfully treats the disease and should therefore be annotated as a negative example of the relation.

The annotators were asked to label all 400 sentences and then re-examine any for which there was disagreement. Inter-annotator agreement (Cohen, 1960) after this stage was of $\kappa = 0.91$ for *may-treat* and $\kappa = 0.94$ for *may-prevent*. Remaining disagreements were resolved by one of the authors based on comments provided by both annotators and the annotation guidelines. The manually annotated version of the data set is referred to as **ML** (Manually Labelled).[4]

[1]http://mbr.nlm.nih.gov

[2]MetaMap annotations use UMLS release 2011AB, http://mbr.nlm.nih.gov/Download/MetaMapped_Medline/

[3]The UMLS's `MRREL` table contains information about related Concept Unique Identifiers (CUIs).

[4]The annotated corpus and further details about the annotation process are available here: https://sites.google.com/site/umlscorpus/home.

	may-prevent						*may-treat*					
	evaluation on DL			evaluation on ML			evaluation on DL			evaluation on ML		
#	prec	rec	f1	prec	rec	f1	prec	rec	f1	prec	rec	f1
2000	33.33	**21.95**	26.47	**44.44**	20.34	**27.91**	44.97	**54.03**	49.08	**48.32**	51.43	**49.83**
4000	27.27	14.63	19.05	**40.91**	**15.25**	**22.22**	46.32	**50.81**	**48.46**	46.32	45.00	45.65
6000	38.89	**17.07**	**23.73**	38.89	11.86	18.18	**54.05**	**64.52**	**58.82**	51.35	54.29	52.78
8000	47.62	**24.39**	**32.26**	**57.14**	20.34	30.00	**57.03**	**58.87**	**57.94**	53.91	49.29	51.49
10000	44.44	**39.02**	41.56	**58.33**	35.59	**44.21**	**61.40**	**56.45**	**58.82**	53.51	43.57	48.03
12000	58.33	**34.15**	**43.08**	58.33	23.73	33.73	**65.05**	**54.03**	**59.03**	53.40	39.29	45.27
14000	**52.38**	**53.66**	**53.01**	50.00	35.59	41.58	**68.89**	**50.00**	**57.94**	57.78	37.14	45.22
16000	**70.83**	**41.46**	**52.31**	58.33	23.73	33.73	**66.02**	**54.84**	**59.91**	55.34	40.71	46.91

Table 2: Results for relation extraction system evaluated against DL and ML data sets

3 Label Comparison

Table 1 shows differences in the annotations for the two techniques for labelling that data. The ML data set for *may-treat* contains 173 positive and 227 negative examples, whereas the ML data set for *may-prevent* contains 139 positives and 261 negatives examples. A comparison of the DL and ML data sets shows that 40.25% of the labels changed for *may-treat* and 39.75% for *may-prevent*. The distant supervision process generated more false positives than false negatives for both relations.

If we assume that we have a classifier that is able to identify the *may-treat* and *may-prevent* relations with perfect accuracy then performance on the ML data sets would be precision=1.0, recall=1.0 and f-score=1.0. However, the false labels on the DL data sets would lead to performance of the same classifiers being estimated as precision=0.61, recall=0.53 and f-score=0.57 for *may-treat* and precision=0.61, recall=0.43 and f-score=0.50 for *may-prevent*. Hence, the two data sets may provide quite different estimates of system performance and we explore this in more detail in the next section.

4 Relation Extraction

A distantly supervised relation classifier was evaluated using manually and distantly labelled versions of the test data. Classifiers were trained for both relations and evaluated using both data sets (DL and ML). The evaluation was carried out using entity level evaluation, i.e. precision and recall are computed based on the proportion of correctly identified entity pairs which occur in sentences labeled as positive examples (according to the annotations contained within DL or ML). Entity level evaluation is commonly used to evaluate distantly supervised relation extraction systems. Similar results have been observed using the alternative approach of sentence level evaluation in which precision and recall are computed by examining the prediction for each sentence.

We use MultiR (Hoffmann et al., 2010), a multi-instance learning system that has been shown to provide state of the art results for distantly supervised relation extraction. The features used are those described by Surdeanu et al. (2011). The system is trained using distantly labelled examples (Section 2.1) of the *may-treat* and *may-prevent* relations containing equal numbers of positive and negative instances. The number of training examples is varied from 2,000 to 16,000 in increments of 2,000.

Results are shown in Table 2. Highlighted figures indicate the data set (DL or ML) against which the highest score was obtained for each metric (prec., rec. and f1) and configuration (relation and number of training examples). In general increasing the amount of training data leads to improved results on the DL data. In particular an increase in precision is observed when there is more training data. However, a different pattern is observed for the ML data and increasing the amount of training data does not always lead to an improvement in the f1-score. Results also show that the performance estimates obtained using the DL and ML data sets are only loosely associated. The results are similar for smaller training data sets but diverge as the amount of training data increases.

The table also shows that for both relations the performance estimates using the DL data are in general higher than those obtained using ML. This trend becomes more pronounced as the amount of training data used increases. The most likely reason for this difference is that the classifiers are trained using distantly supervised data and therefore model the labels in the DL data set more closely than those in found in ML.

These results demonstrate that evaluation using distantly labelled gold standard data tends to overestimate performance. In some cases the discrepancy is large (up to 18.58 for *may-prevent* and 13.76 for *may-treat*). However, it does not seem to be consistent or particularly predictable. Consequently, improving the performance of a relation extraction system relative to distantly labelled evaluation data does not necessarily imply an increase in performance when measured against a manually annotated gold-standard.

5 Conclusion

This paper explored the effect of evaluating biomedical relation extraction systems using held-out test data annotated using distant supervision. Test data for two biomedical relations was annotated using distant supervision and also manually annotated. The manual and automatic labels differed for a large portion of the sentences. A distantly supervised relation extraction system was also evaluated using both data sets. We found that evaluation using held-out distantly supervised data tended to overestimate performance and that the connection between improved performance against distantly and manually labelled data was unclear. The use of held-out distantly labelled data is a cheap and efficient way to evaluate relation extraction systems, however this analysis demonstrates that the results obtained should be treated with some caution and, ideally, systems should also be evaluated against manually labelled data.

The results presented here were obtained for two biomedical relations. In future we plan to extend our analysis to a wider set of relations.

Acknowledgments

The authors are grateful to the Engineering and Physical Sciences Research Council (EP/J008427/1) for funding the research described in this paper.

References

A. Aronson and F. Lang. 2010. An overview of MetaMap: historical perspective and recent advances. *Journal of the American Medical Association*, 17(3):229–236.

Isabelle Augenstein, Diana Maynard, and Fabio Ciravegna. 2014. Relation extraction from the web using distant supervision. In *Proceedings of the 19th International Conference on Knowledge Engineering and Knowledge Management (EKAW 2014)*, Linköping, Sweden, November.

Kedar Bellare and Andrew Mccallum. 2007. Learning Extractors from Unlabeled Text using Relevant Databases. In *Sixth International Workshop on Information Integration on the Web (IIWeb)*.

Jacob Cohen. 1960. A coefficient of agreement for nominal scales. *Educational and Psychological Measurement*, 20:37–46.

Mark Craven and Johan Kumlien. 1999. Constructing biological knowledge bases by extracting information from text sources. In *In Proceedings of the Seventh International Conference on Intelligent Systems for Molecular Biology (ISMB)*, pages 77–86. AAAI Press.

Raphael Hoffmann, Congle Zhang, and Daniel S. Weld. 2010. Learning 5000 relational extractors. In *Proceedings of the 48th Annual Meeting of the Association for Computational Linguistics*, ACL '10, pages 286–295, Stroudsburg, PA, USA. Association for Computational Linguistics.

Ander Intxaurrondo, Mihai Surdeanu, Oier Lopez de Lacalle, and Eneko Agirre. 2013. Removing noisy mentions for distant supervision. *Procesamiento del Lenguaje Natural*, 51:41–48.

Sebastian Krause, Hong Li, Hans Uszkoreit, and Feiyu Xu. 2012. Large-scale learning of relation-extraction rules with distant supervision from the web. In *Proceedings of the 11th International Conference on The Semantic Web - Volume Part I*, ISWC'12, pages 263–278, Berlin, Heidelberg. Springer-Verlag.

Bonan Min, Ralph Grishman, Li Wan, Chang Wang, and David Gondek. 2013. Distant supervision for relation extraction with an incomplete knowledge base. In *Proceedings of the 2013 Conference of the North American Chapter of the Association for Computational Linguistics: Human Language Technologies*, pages 777–782, Atlanta, Georgia, June. Association for Computational Linguistics.

Mike Mintz, Steven Bills, Rion Snow, and Dan Jurafsky. 2009. Distant supervision for relation extraction without labeled data. In *Proceedings of the Joint Conference of the 47th Annual Meeting of the ACL and the 4th International Joint Conference on Natural Language Processing of the AFNLP: Volume 2 - Volume 2*, ACL '09, pages 1003–1011,

Stroudsburg, PA, USA. Association for Computational Linguistics.

Claire Nédellec, Robert Bossy, Jin-Dong Kim, Jung-Jae Kim, Tomoko Ohta, Sampo Pyysalo, and Pierre Zweigenbaum. 2013. Overview of bionlp shared task 2013. In *Proceedings of the BioNLP Shared Task 2013 Workshop*, pages 1–7, Sofia, Bulgaria, August. Association for Computational Linguistics.

Truc-Vien T. Nguyen and Alessandro Moschitti. 2011. Joint distant and direct supervision for relation extraction. In *Proceedings of The 5th International Joint Conference on Natural Language Processing (IJCNLP'11)*, pages 732–740. Association for Computational Linguistics.

Sebastian Riedel, Limin Yao, and Andrew McCallum. 2010. Modeling relations and their mentions without labeled text. In *Proceedings of the European Conference on Machine Learning and Knowledge Discovery in Databases (ECML PKDD '10)*.

Alan Ritter, Luke Zettlemoyer, Mausam, and Oren Etzioni. 2013. Modeling missing data in distant supervision for information extraction. In *Association for Computational Linguistics Vol. 1 (TACL)*.

Roland Roller and Mark Stevenson. 2014. Applying umls for distantly supervised relation detection. In *Proceedings of the Fifth International Workshop on Health Text Mining and Information Analysis (Louhi 2014)*, Gothenburg, Sweden.

Roland Roller, Eneko Agirre, Aitor Soroa, and Mark Stevenson. 2015. Improving distant supervision using inference learning. In *Proceedings of the 53rd Annual Meeting of the Association for Computational Linguistics and the 7th International Joint Conference on Natural Language Processing (Volume 2: Short Papers)*, pages 273–278, Beijing, China, July. Association for Computational Linguistics.

Benjamin Roth and Dietrich Klakow. 2013. Combining generative and discriminative model scores for distant supervision. In *Proceedings of the 2013 Conference on Empirical Methods in Natural Language Processing*, pages 24–29, Seattle, Washington, USA, October. Association for Computational Linguistics.

Isabel Segura-Bedmar, Paloma Martínez, and María Herrero Zazo. 2013. Semeval-2013 task 9 : Extraction of drug-drug interactions from biomedical texts (ddiextraction 2013). In *Second Joint Conference on Lexical and Computational Semantics (*SEM), Volume 2: Proceedings of the Seventh International Workshop on Semantic Evaluation (SemEval 2013)*, pages 341–350, Atlanta, Georgia, USA, June. Association for Computational Linguistics.

Mihai Surdeanu, David McClosky, Mason Smith, Andrey Gusev, and Christopher Manning. 2011. Customizing an information extraction system to a new domain. In *Proceedings of the ACL 2011 Workshop on Relational Models of Semantics*, pages 2–10, Portland, Oregon, USA, June. Association for Computational Linguistics.

Mihai Surdeanu, Julie Tibshirani, Ramesh Nallapati, and Christopher D. Manning. 2012. Multi-instance multi-label learning for relation extraction. In *Proceedings of the 2012 Joint Conference on Empirical Methods in Natural Language Processing and Computational Natural Language Learning*, EMNLP-CoNLL '12, pages 455–465, Stroudsburg, PA, USA. Association for Computational Linguistics.

Shingo Takamatsu, Issei Sato, and Hiroshi Nakagawa. 2012. Reducing wrong labels in distant supervision for relation extraction. In *Proceedings of the 50th Annual Meeting of the Association for Computational Linguistics: Long Papers - Volume 1*, ACL '12, pages 721–729, Stroudsburg, PA, USA. Association for Computational Linguistics.

Philippe Thomas, Illés Solt, Roman Klinger, and Ulf Leser. 2011. Learning protein protein interaction extraction using distant supervision. In *Proceedings of Robust Unsupervised and Semi-Supervised Methods in Natural Language Processing*, pages 34–41.

Andreas Vlachos and Stephen Clark. 2014. Application-driven relation extraction with limited distant supervision. In *Proceedings of the First AHA!-Workshop on Information Discovery in Text*, pages 1–6, Dublin, Ireland, August. Association for Computational Linguistics and Dublin City University.

Fei Wu and Daniel S. Weld. 2007. Autonomously semantifying wikipedia. In *Proceedings of the Sixteenth ACM Conference on Conference on Information and Knowledge Management*, CIKM '07, pages 41–50, New York, NY, USA. ACM.

Wei Xu, Raphael Hoffmann, Le Zhao, and Ralph Grishman. 2013. Filling knowledge base gaps for distant supervision of relation extraction. In *Proceedings of the 51st Annual Meeting of the Association for Computational Linguistics (Volume 2: Short Papers)*, pages 665–670, Sofia, Bulgaria, August. Association for Computational Linguistics.

Xingxing Zhang, Jianwen Zhang, Junyu Zeng, Jun Yan, Zheng Chen, and Zhifang Sui. 2013. Towards accurate distant supervision for relational facts extraction. In *Proceedings of the 51st Annual Meeting of the Association for Computational Linguistics (Volume 2: Short Papers)*, pages 810–815, Sofia, Bulgaria, August. Association for Computational Linguistics.

Checking a structured pathology report for completeness of content using terminological knowledge

Sebastian Busse
Department Informatics and Media
University of Applied Sciences
Brandenburg, Germany
busses@fh-brandenburg.de

Abstract

Structuring of information helps people to gain a quick overview of complex issues and facilitates the transfer of large amounts of data. In the medical field, such data are transferred using defined standards (HL7[1], DICOM[2]) or in conjunction with terminology systems (ICD-10[3], LOINC[4], SNOMED CT[5]). This paper focuses on the structuring of diagnostic reports in the field of anatomic pathology. It describes how to make the content of these reports semantically understandable for machines. Finally, it will be shown that structured pathology reports can be checked for completeness of content in a computerized way by using terminological knowledge. For this purpose, an ontology has been designed that describes the subdomain of reporting a radical prostatectomy specimen.

1 Introduction

The advantage of a structured report against an unstructured free text is that it can be divided into subareas with definable context. For each disease occurring in the field of pathology, it can be determined how to investigate it and how to structure and encode the description of the examination results. Supporting the pathologist in documenting his observations, could help to avoid missing data in the report.

In conjunction with medical terminologies, a suitable report structure improves the addressability of particular contents for machines. There are different approaches on mapping the clinical terms occurring in free texts of documentations to medical terminologies, such as SNOMED CT, by using text mining methods (Stenzhorn et al., 2009; Spasic et al., 2005; Allones et al., 2014). Extracting machine-readable facts out of raw text facilitates the electronic exchange of the report information between information technology (IT) systems (Bouhaddou et al., 2008; White and Carolan-Rees, 2013). Moreover, the structuring of reports allows a software-controlled search for defined elements. This simplifies searching stored reports for specific study criteria or diagnoses (Brown and Soenksen, 2010).

Currently, such a workflow seems not to be feasible in practice. The problem is that existing medical terminologies do not adequately contain all the observations and specimen collection procedures that are required to be available for the pathology domain (Daniel et al., 2011).

This paper describes how terminological knowledge covering the scope of reporting a radical prostatectomy specimen can be arranged for the purpose of checking particular pathology reports for their completeness of content.

2 Materials and methods

2.1 The pathology structured report

The IHE[6] Anatomic Pathology working group created a technical framework that contains the specification of Anatomic Pathology Structured Reports (APSR) (Daniel and Macary, 2011). This specification defines the APSR content profile, which is the result of a joint initiative from IHE and HL7 Anatomic Pathology working groups.

[1]Health Level Seven: http://www.hl7.org

[2]Digital Imaging and Communications in Medicine: http://medical.nema.org

[3]International Statistical Classification of Diseases and Related Health Problems 10th Revision: http://apps.who.int/classifications/icd10/browse/2015/en

[4]Logical Observation Identifiers Names and Codes: https://loinc.org

[5]Systematized Nomenclature Of Medicine Clinical Terms: http://www.ihtsdo.org/snomed-ct

[6]Integrating the Healthcare Enterprise: http://www.ihe.net

Proceedings of the Sixth International Workshop on Health Text Mining and Information Analysis (Louhi), pages 103–108, Lisbon, Portugal, 17 September 2015.

Furthermore, it serves as a trial implementation describing the realization of the APSR content profile using the HL7 Clinical Document Architecture (CDA) (Dolin et al., 2006).

Such a CDA-based APSR basically consists of a header and a body. The header contains information about the context of the treatment order, the patient data and the examining pathologist. The body contains various hierarchical structured sections. Each section describes its content in the form of human-readable text. In addition, some sections contain entry elements, which convert the human-readable information from the text element in machine-readable data. Therefore, each entry element references a particular concept, which is described semantically within a terminology. The address that references a concept is called URI (uniform resource identifier). According to these specifications, an APSR contains both human- and machine-readable information.

This way, an APSR references its content to concepts of terminologies. That has the advantage of being able to identify a specific content by a unique URI in every report and hence give a semantic meaning to this content.

2.2 The terminological knowledge base

Terminologies help to structure concepts of a specific subject area in a certain language by using a common vocabulary that is as consensual as possible. (Roche et al., 2009)

The aim of checking a report for its completeness of content includes the need to determine what content is required. The CAP (College of American Pathologists) offers some cancer protocols[7], which specify the content of pathology reports for different cancer types. Moreover, the ICCR group (International Collaboration on Cancer Reporting) has published five datasets[8] for reporting different types of cancer. These determine which information is required in a report and which information is just considered to be recommended.

Daniel and Macary (2011) created a terminology called PathLex[9], which covers the scope of anatomic pathology observations and specimen collection procedures. The aim of PathLex is to achieve semantic consistency of standard messages and document structures within and across standards (HL7, DICOM). That means to guarantee that various information systems create equally structured clinical information which are both human- and machine-readable. Therefore, a unified knowledge base is needed that adopts the knowledge of existing terminology systems - such as SNOMED CT and ICD-O[10] - and fills critical knowledge gaps using newly defined concepts.

PathLex is an "interface terminology" (Daniel et al., 2011). In clinical settings, such terminologies support clinicians in entering information into computer programs by providing a systematic collection of clinically oriented phrases (terms such as "Gleason Score" or "Margin status"). In the opposite way, interface terminologies facilitate the presentation of electronically stored, machine-readable patient information as human-readable text that the clinician can read easier (Rosenbloom et al., 2006). Accordingly, PathLex provides a range of flexible "pathologist friendly" phrases, but raises no claim to be a complete, all-encompassing semantic representation for the contained concepts in relation to the entire medical knowledge in reality.

As an interface terminology, the strategy of PathLex regarding the semantic interoperability is to derive concepts out of the phrases used by pathologists and then linking them to reference terminologies. Mapping interface terminologies to standard reference terminologies rather than identifying one or more interface terminologies to serve as standards is a commonly admitted strategy towards semantic interoperability (Rosenbloom et al., 2009). Newly defined concepts that do not appear in any reference terminology so far must be explained with the aid of known concepts and relations. The known concepts are linked to their representations in existing reference terminologies. In this way, PathLex could comprehensively represent the knowledge base of the anatomic pathology domain and serve as an aid in semantically structuring regarding the creation process of pathology reports. Currently, the mapping of PathLex concepts to the reference terminology SNOMED CT is solely realized by an algorithm of the National Center for Biomedi-

[7]CAP cancer protocols: http://www.cap.org/web/home/resources/cancer-reporting-tools/cancer-protocol-templates

[8]ICCR cancer datasets: http://www.iccr-cancer.org/datasets

[9]PathLex - OID : 1.3.6.1.4.1.19376.1.8.2.1, http://bioportal.bioontology.org/ontologies/PATHLEX

[10]International Classification of Diseases for Oncology: http://www.who.int/classifications/icd/adaptations/oncology/en

cal Ontology (NCBO) called LOOM, which automatically relates two terms based on close lexical match between their preferred names or the preferred name of a term and a defined synonym of another. The lexical match involves removing white-space and punctuation from the considered labels. Due to the existing concepts in SNOMED CT, that has the effect that the mapping is well advanced for some pathological observations (for example, in the area of histological observations), whereas it can not possibly exist for others where there are no predefined concepts with the required lexical match, let alone the corresponding meaning, available in the reference terminology (for example, the TNM classification of tumors) (Daniel and Macary, 2011). According to the BioPortal website[11], the LOOM algorithm generated 340 mappings from PathLex concepts to SNOMED CT concepts. The APSR content profile uses PathLex to encode textual observations in order to define templates for sharing and exchanging the reports (Daniel et al., 2012).

2.3 Methods

In order to obtain the ability of checking whether a pathology report is complete in terms of content, it is necessary to determine the required contents. Therefore, the ICCR prostate cancer dataset[12] was used to identify the contents that are required and the ones that are considered to be recommended.

The next step was to structure the report and find a possibility to reference its contents to the concepts of terminology systems. In this paper, the IHE Anatomic Pathology CDA-based APSR structure was used to construct five invented example reports[13] with the properties as shown in Table 1.

In order to check these reports for completeness of content in a computerized way, it was necessary to describe the content requirements in machine-readable code. That can be achieved by using a suitable terminology.

As explained in Section 3, PathLex could not be used to describe the desired properties semantically correct. For this reason, the content of

Structured report	Missing required contents	Missing recommended contents
Example 1	0	0
Example 2	0	4
Example 3	2	0
Example 4	4	2
Example 5	All (17)	All (6)

Table 1: Properties regarding the completeness of content of the five constructed example reports.

the ICCR prostate cancer dataset was used to develop an adapted ontology that was named PathLexProstate[14] and can be seen as a terminological knowledge base containing the concepts of the dataset. PathLexProstate was created using the free, open-source ontology editor Protégé[15] and is saved in the functional-style syntax of the Web Ontology Language (OWL) 2 as defined by Motik et al. (2009).

The contents contained in the entry elements of the five example reports were linked to the appropriate concepts of PathLexProstate.

2.4 Evaluation procedure

Finally, an evaluation procedure was designed to check the example reports for completeness of content. Therefore, the contents of an example report and the concepts of the PathLexProstate knowledge base are read in. The ontology specifies which contents are required and which ones are recommended, whereas the entry elements of the CDA-based report state which contents are included. The evaluation procedure compares these inputs and then draws a conclusion about the report completeness in terms of content. As defined in the ICCR dataset and described by Kench et al. (2013), a report does not need to contain any recommended content to be counted as complete, but at least it has to contain all the required contents.

3 Results

In order to check pathology reports for completeness of content, it is initially necessary to structure the human-readable free text of these reports

[11] Mappings of PathLex concepts: http://bioportal.bioontology.org/ontologies/PATHLEX/?p=mappings

[12] ICCR Prostate Cancer Dataset: http://www.iccr-cancer.org/datasets/published-datasets/urinary-male-genital/prostate-cancer-radical-prostatectomy-specimen

[13] Pathology report examples: http://sourceforge.net/projects/pathlexprostate/files/PathologyReportExamples

[14] PathLexProstate v1.0: http://sourceforge.net/projects/pathlexprostate/files/PathLexProstate

[15] Protégé: http://protege.stanford.edu

into sections, which can then be addressed by machines. The IHE Anatomic Pathology APSR content profile describes a possible variant of such a structuring. In the specified CDA-based documents, the contents which are included in the text element of each section are defined in the particular entry elements by referencing them to concepts described in terminology systems.

The analysis of PathLex has revealed that this interface terminology is not ready to be used as a part of the desired pathology report conformance check as far as their completeness in terms of content is considered. Although the mapping of PathLex concepts to the reference terminology SNOMED CT is already performed 340 times, there are structural issues that could lead to semantically wrong interpretations of some concepts. PathLex does not contain any Properties. This means that the relationships between the defined classes are not shown. The only exception is the default is-a-relationship, which defines a class as a subclass of another. However, these is-a-relations are not always semantically correct. Consequently, there can be no hierarchical classification of the concepts contained in PathLex.

Moreover, a representation of a pathology report needs to be added to the knowledge base. That has the advantage that this representation can then be related to the concepts which are representing the particular required report contents.

For these reasons, the ontology PathLexProstate was created. Figure 1 shows the class named *"ICCR Prostate Cancer Report"* in the center of the image, which represents the concept of pathology reports as specified by the ICCR prostate cancer dataset. The solid line with the arrow in the direction of the ICCR report class displays that this is a subclass of the class *"Pathology report"*. This expresses that every single ICCR prostate cancer report is a pathology report. The broken lines display the relations of the ICCR report concept with the concepts of the desired report contents. According to the ICCR prostate cancer dataset, there are 17 required (dark gray broken lines) and 6 recommended (light gray broken lines) classes surrounding the center of the image. In total, PathLexProstate contains 118 classes and two object properties (*"Contains required information about"* and *"Contains recommended information about"*) besides the default subclass relationship.

Using PathLexProstate and the evaluation procedure described in Subsection 2.4, the completeness check could be carried out correctly for all the five example reports as they were previously specified (see Table 1). As expected, a lack of required contents always led to a negative test result by displaying the missing concepts as errors and stating that the considered report is not complete in terms of content, whereas the presence or absence of recommended contents had no effect on the result of the completeness check. Nevertheless, the absence of a recommended concept was correctly displayed as a warning in any case. Serving as a proof, Figure 2 shows the result of the evaluation procedure of the report example 2. As stated in Table 1, this report includes all the required concepts, whereas four recommended ones are missing. In conclusion, this report is correctly detected as being "complete in terms of content".

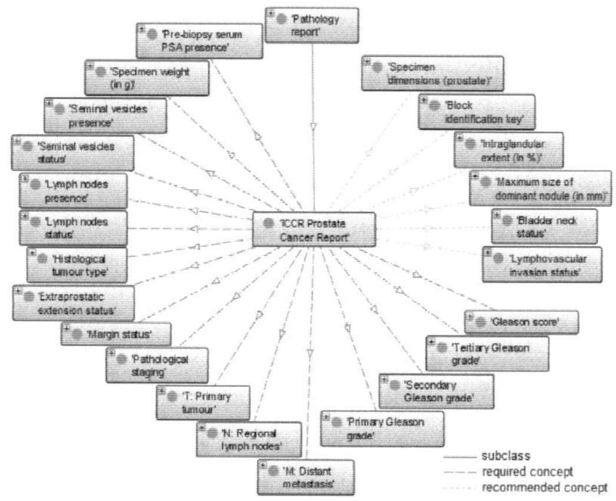

Figure 1: Representation of the ICCR prostate cancer report in PathLexProstate

```
Result:
The report 'Pathology_Report_Example2.xml'
is complete in terms of content.
Errors:
0 required concepts missing.
Warnings:
4 recommended concepts missing.
000054: Block identification key
000112: Lymphovascular invasion status
000161: Intraglandular extent (in %)
000170: Maximum size of dominant nodule (in mm)
```

Figure 2: Result of the evaluation procedure regarding the report example 2.

4 Discussion

In summary, it can be said that the developed ontology PathLexProstate can be used as terminological knowledge base for checking pathology prostate cancer reports for completeness in terms of content according to the ICCR prostate cancer dataset.

The described method needs a CDA-based structured report and a suitable terminological knowledge base to perform the evaluation procedure.

Working with the IHE APSR content profile for structuring a pathology report and then referencing its contained contents to terminologies as described in Subsection 2.1, offers the possibility to gain machine-readable reports.

The terminological knowledge needs to contain the classes that represent the required report contents. Additionally, the specific report has to be defined as a class and its relations to the needed report contents must be specified. Consequently, the development of such a knowledge base requires initially the help of domain experts, who have to determine the contents that have to be included in a complete report. The problem is that there are many different guidelines determining report requirements and the majority of them do not serve as worldwide standard. The CAP and the ICCR formed cancer report templates, which are already internationally accepted (Srigley et al., 2009; Baskovich and Allan, 2011; Kench et al., 2013). For this reason, the ICCR prostate cancer dataset was chosen to determine the minimum dataset of pathology reports in this field and then create the terminological knowledge base PathLexProstate. Including more organizations while defining report templates, could lead to worldwide acceptance and consistent minimum datasets for the future.

Currently, medical terminologies do not contain any information about report contents that are considered as recommended or even required for completeness of content. PathLexProstate tries to offer an example on coding these determinations for the scope of pathology reports of radical prostatectomy specimens.

The described report conformance check should be seen as a supporting method and not as a barrier that strictly forces any content in pathology reports. It can be used to help pathologists during the process of documenting their observations by mentioning potential content-related gaps in the report in order to avoid missing data. Nevertheless, it should not forbid writing or saving a pathology report, even if it is detected as being incomplete. The conformance check is just planned to warn the clinician if useful data could have been forgotten to enter.

Moreover, the conformance check could be used for filtering existing pathology reports based on content-related requirements. This can be interesting for re-use purposes, such as scientific studies, in the future.

Although the described method can check for completeness, the semantic plausibility of report contents has not been verified so far. The developed ontology should be seen as an interface terminology. Mapping the contained classes to a reference terminology, such as SNOMED CT, could help extending the semantic expressiveness of the concepts covered in PathLexProstate.

References

Jose L. Allones, Diego Martinez and Maria Taboada. (2014). *Automated Mapping of Clinical Terms into SNOMED-CT. An Application to Codify Procedures in Pathology*. J Med Syst. 2014 Oct; 38(10): 134.

Brett W. Baskovich and Robert W. Allan. (2011). *Web-based synoptic reporting for cancer checklists*. J Pathol Inform. 2011 Mar; 2: 16.

Omar Bouhaddou, Pradnya Warnekar, Fola Parrish, Nhan Do, Jack Mandel, John Kilbourne and Michael J. Lincoln. (2008). *Exchange of Computable Patient Data between the Department of Veterans Affairs (VA) and the Department of Defense (DoD): Terminology Mediation Strategy*. J Am Med Inform Assoc. 2008 Mar-Apr; 15(2): 174-183.

Philip J. B. Brown and Peter Soenksen. (2010). *Evaluation of the Quality of Information Retrieval of Clinical Findings from a Computerized Patient Database Using a Semantic Terminological Model*. J Am Med Inform Assoc. 2000 Jul-Aug; 7(4): 392-403.

Christel Daniel and Francois Macary. (2011). *IHE Anatomic Pathology Technical Framework Supplement - Anatomic Pathology Structured Reports (APSR) - Trial Implementation*, Rev. 1.1. http://www.ihe.net/Technical_Framework/upload/IHE_PAT_Suppl_APSR_Rev1-1_TI_2011_03_31.pdf.

Christel Daniel, Francois Macary, Marcial Garcia Rojo, Jacques Klossa, Arvydas Laurinavicius, Bruce A. Beckwith and Vincenzo Della Mea. (2011). *Recent advances in standards for collaborative Digital Anatomic Pathology*. Diagn Pathol. 2011 Mar 30; 6 Suppl 1: S17.

Christel Daniel, David Booker, Bruce Beckwith, Vincenzo Della Mea, Marcial Garcia-Rojo, Lori Havener, Mary Kennedy, Jacques Klossa, Arvydas Laurinavicius, Francois Macary, Vytenis Punys, Wendy Scharber and Thomas Schrader. (2012). *Standards and Specifications in Pathology: Image Management, Report Management and Terminology.* Stud Health Technol Inform. 2012; 179: 105-122.

Robert H. Dolin, Liora Alschuler, Sandy Boyer, Calvin Beebe, Fred M. Behlen, Paul V. Biron and Amnon Shabo (Shvo). (2006). *HL7 Clinical Document Architecture, Release 2.* J Am Med Inform Assoc. 2006 Jan-Feb; 13(1): 30-39.

James G Kench, Brett Delahunt, David F Griffiths, Peter A Humphrey, Thomas McGowan, Kiril Trpkov, Murali Varma, Thomas M Wheeler and John R Srigley. (2013). *Dataset for reporting of prostate carcinoma in radical prostatectomy specimens: recommendations from the International Collaboration on Cancer Reporting.* Histopathology. 2013 Jan; 62(2): 203-218.

Boris Motik, Peter F. Patel-Schneider, Bijan Parsia, Conrad Bock, Achille Fokoue, Peter Haase, Rinke Hoekstra, Ian Horrocks, Alan Ruttenberg, Uli Sattler and Mike Smith. (2009). *OWL 2 Web Ontology Language: Structural Specification and Functional-Style Syntax.* `http://www.w3.org/2009/pdf/REC-owl2-syntax-20091027.pdf`.

Christophe Roche, Marie Calberg-Challot, Luc Damas and Philippe Rouard. (2009). *Ontoterminology: A new paradigm for terminology.* International Conference on Knowledge Engineering and Ontology Development, 2009 Oct, pp.321-326.

S. Trent Rosenbloom, Randolph A. Miller, Kevin B. Johnson, Peter L. Elkin and Steven H. Brown. (2006). *Interface Terminologies: Facilitating Direct Entry of Clinical Data into Electronic Health Record Systems.* J Am Med Inform Assoc. 2006 May-Jun; 13(3): 277-288.

S. Trent Rosenbloom, Steven H. Brown, David Froehling, Brent A. Bauer, Dietlind L. Wahner-Roedler, William M. Gregg and Peter L. Elkin. (2009). *Using SNOMED CT to Represent Two Interface Terminologies.* J Am Med Inform Assoc. 2009 Jan-Feb; 16(1): 81-88.

Irena Spasic, Sophia Ananiadou, John McNaught and Anand Kumar. (2005). *Text mining and ontologies in biomedicine: Making sense of raw text.* Brief Bioinform. 2005 Sep; 6(3): 239-251.

John R. Srigley, Tom McGowan, Andrea MacLean, Marilyn Raby, Jillian Ross, Sarah Kramer and Carol Sawka. (2009). *Standardized synoptic cancer pathology reporting: A population-based approach.* J Surg Oncol. 2009 Jun 15; 99(8): 517-524.

Holger Stenzhorn, Edson Jose Pacheco, Percy Nohama and Stefan Schulz. (2009). *Automatic Mapping of Clinical Documentation to SNOMED CT.* Stud Health Technol Inform. 2009; 150: 228-232.

Judith White and Grace Carolan-Rees. (2013). *Current state of medical device nomenclature and taxonomy systems in the UK: spotlight on GMDN and SNOMED CT.* JRSM Short Rep. 2013 Jun 5; 4(7): 1-7.

Effectively Crowdsourcing Radiology Report Annotations

Anne Cocos[1,2], Aaron J. Masino[1], Ting Qian[1], Ellie Pavlick[2], and Chris Callison-Burch[2]
Department of Biomedical and Health Informatics, The Children's Hospital of Philadelphia[1]
Computer and Information Sciences Department, University of Pennsylvania[2]

Abstract

Crowdsourcing platforms are a popular choice for researchers to gather text annotations quickly at scale. We investigate whether crowdsourced annotations are useful when the labeling task requires medical domain knowledge. Comparing a sentence classification model trained with expert-annotated sentences to the same model trained on crowd-labeled sentences, we find the crowdsourced training data to be just as effective as the manually produced dataset. We can improve the accuracy of the crowd-fueled model without collecting further labels by filtering out worker labels applied with low confidence.

1 Introduction

Most text classification methods are based on supervised machine learning models that require large amounts of labeled training data (Aggarwal and Zhai, 2012). Gathering a large amount of high-quality training data can be time-consuming and expensive. To streamline the process, natural language processing (NLP) researchers have employed crowdsourcing platforms to quickly collect crowdsourced annotations at scale (Khare et al., 2015; Wang et al., 2013).

In some NLP problems, the annotation task requires some degree of common linguistic knowledge that most non-experts are assumed to have. By examining the accuracy of crowdsourced data and its usefulness in training models to perform common NLP tasks, previous research has shown that deficiencies in individual crowd worker accuracy can be overcome by taking consensus votes over multiple annotators or weighting the votes of annotators based on their overall performance (MacLean and Heer, 2013; Zhai et al., 2013; Hsueh et al., 2009; Snow et al., 2008).

But how useful is crowdsourcing when the annotation task requires domain knowledge beyond common knowledge? One example is interpretation of medical data. As hospitals transition to electronic patient records, there are increasingly more data than medical experts have time to manually annotate. If crowdsourced medical annotations prove to be mostly accurate, it will accelerate research on using machine learning methods to support medical decisions.

Previous research has suggested that crowdsourced non-experts are capable of identifying distinct patterns of activity in electroencephalography readings (Warby et al., 2014) and predicting native protein structures (Cooper et al., 2010). To our knowledge there has been less work in using unscreened, crowdsourced workers to complete text labeling tasks that require comprehension of medical concepts. Consider the task of determining whether these excerpts from a radiology report describe a *normal* or *abnormal* observation of the anatomical structure in parentheses:[1]

- *The mastoid air cells are well-pneumatized.* (mastoid)
- *Bilateral dysplastic vestibules and lateral semicircular canals.* (semicircular canal)
- *The external auditory canal is patent.* (EAC)

Labeling some of these sentences might require a non-expert to do additional research. (e.g. Should a mastoid air cell be pneumatized? Does *lateral* describe the condition of the semicircular canal, or is *lateral semicircular canal* a compound noun?) In this work, we extend the study of crowdsourcing annotations to text-labeling tasks that require domain knowledge. Specifically, we examine the usefulness of crowdsourced data for training models to classify radiology report sentences as *normal* or *abnormal* as in the examples above. By

[1] The true labels are [*normal, abnormal, normal*].

Proceedings of the Sixth International Workshop on Health Text Mining and Information Analysis (Louhi), pages 109–114,
Lisbon, Portugal, 17 September 2015.

comparing the performance of classification models trained on expert-generated and crowdsourced data sets, we show that crowdsourcing enables us to build supervised models without sacrificing accuracy. Additionally, we show that as gains in accuracy achieved by increasing the training set size level off, we can further improve the accuracy of our classifier – without gathering additional training data – by incorporating worker confidence votes.

2 Methods and Data Collection

2.1 Annotating radiology report reports

The Audiological and Genetic Database (AudGenDB) (CHOP, 06) is a medical research database that houses over 16,000 radiology images of the temporal bones and associated text reports. The reports are unlabeled, making it difficult for researchers to filter reports containing abnormalities in a particular component of the ear. The motivation for our work is to build a model that classifies each report as *normal* or *abnormal* with respect to each of thirteen anatomical structures (e.g. cochlea, stapes, etc.). Here, we focus specifically on the sub-task of classifying sentences in a report as *normal* or *abnormal*.

2.2 Data collection

Our full data set consists of 10,880 unlabeled sentences extracted from AudGenDB radiology reports, similar to the examples in the introduction and in the supplemental material.

2.2.1 Gold standard labels: expert annotations

Two experts individually annotated a randomly chosen sample of 340 sentences. The experts achieved an inter-annotator agreement score of 0.848 (Fleiss Kappa/Krippendorffs Alpha), indicating near-perfect agreement (Landis and Koch, 1977) . The final gold standard dataset includes only the 323 sentences on which both annotators agreed on the label: 165 (51.1%) *normal* and 158 (48.9%) *abnormal*.

2.2.2 Crowdsourced annotations

We collected crowdsourced annotations using the Amazon Mechanical Turk (MTurk) crowdsourcing platform. To facilitate annotation we created an interface to show each worker three sentences per Human Intelligence Task (HIT). We performed no screening of the workers for medical or radiology expertise, and assumed them to be non-experts.

To encourage high quality annotations we provided workers with brief instructions to *"classify the highlighted sentence as describing a normal or abnormal observation of the specified ear component"* and examples of *normal* and *abnormal* sentences (figure 1). This was the only training provided. We monitored performance on each HIT using at least one control sentence from the gold standard dataset with known class.

In addition to asking the workers to indicate whether each sentence described a *normal* or *abnormal* observation, we also asked them to indicate their confidence (*Very Confident*, *Somewhat Confident*, or *Not Confident*) in their choice, serving as a self-reported measure of either the perceived difficulty of classifying particular sentences, the accuracy of their classifications, or both.

We solicited labels for each unlabeled sentence from at least two unique workers. If workers disagreed on a sentence label we continued to collect annotations until reaching 75% absolute agreement. In total, we collected annotations satisfying these conditions for 717 additional sentences employing 56 unique workers. Data collection took under two days and cost less than $600 USD.

2.2.3 Weighting the workers' votes

To consolidate MTurk workers individual votes into a single crowdsourced label for each sentence, we take the class of each sentence as the weighted average of the workers votes. Following Snow et al. (2008), we weight the workers votes based on their accuracy. Intuitively, we weigh the votes of accurate workers higher than votes of inaccurate workers. Further, if two workers achieve the same percentage accuracy over a different number of sentences, we want to weigh more heavily the votes of the worker who classified more sentences. To satisfy these criteria, we employed the lower bound of the 25% Clopper-Pearson binomial confidence interval for worker accuracy as a heuristic weighting scheme:

$$w_i = B(\frac{1 - 0.25}{2}; C, T - C + 1)$$

where w_i is the weight given to the annotations from worker i, B is the beta distribution probability density function, T is the total number of con-

Component	*Class*	*Example Text*
scutum	normal	There is no evidence of bony erosion of the ossicles or the scutum.
ossicles	abnormal	The ossicles are markedly dysplastic.
auditory canal	normal	The internal auditory canal is unremarkable.
stapes	abnormal	The stapes is thickened.

Figure 1: Sample of example sentences provided to workers

trol sentences annotated by i, and C is the number of control sentences correctly annotated by i.

2.3 Building a classification model

We constructed a simple sentence classification model using a bag-of-ngrams sentence representation to examine whether the crowdsourced data is as useful for training a sentence classification model as the expert-annotated data.

Our model represents each tokenized sentence as a 901-dimensional feature vector, where the first 900 features correspond to the top-500 unigrams, top-300 bigrams, and top-100 trigrams in our dataset in terms of frequency. The value of each n-gram feature indicates the count of that n-gram in the sentence. The 901st feature indicates the sentence token count. Having represented each sentence as a numeric feature vector, we use L2-regularized logistic regression to predict whether the sentence is *normal* or *abnormal*.

3 Results

3.1 Labeling performance and analysis

Our 56 unique MTurk workers each classified 99.9 sentences on average (range [3, 462]). The average individual accuracy on classifying control sentences was 93.49%, and performance was relatively consistent between workers. Only three workers had accuracy scores significantly below average as determined by the 95% binomial proportion confidence interval.

Similarly to previous studies that examine the reliability of crowdsourced annotations (Zhai et al., 2013; Hsueh et al., 2009; Snow et al., 2008), we find that inter-annotator agreement among the crowdsourced workers was lower than agreement between our expert annotators. We calculate inter-annotator agreement using two methods. Applying Krippendorffs Alpha directly, the crowdsourced workers achieve a score of 0.743. Because a varying number of workers labeled each crowdsourced sentence, we cannot calculate Fleiss Kappa directly as we could for the two expert annotators. Instead we randomly sample two crowd labels for each sentence for 100 iterations and find the average Kappa score over all iterations to be 0.758 (90% CI $\pm$.003). This indicates substantial agreement (Landis and Koch, 1977), albeit lower than agreement between the expert annotators who scored 0.848 on both measures.

3.2 Votes of confidence

Workers generally indicated high confidence in their annotations. The distribution of ratings was 68% *Very Confident*, 27% *Somewhat Confident*, and 5% *Not Confident*.

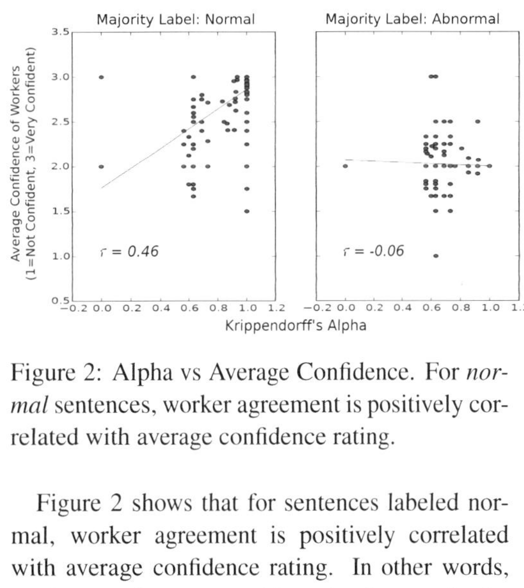

Figure 2: Alpha vs Average Confidence. For *normal* sentences, worker agreement is positively correlated with average confidence rating.

Figure 2 shows that for sentences labeled normal, worker agreement is positively correlated with average confidence rating. In other words, workers tend to agree with each other on the labeling of a sentence when they each feel confident in their own judgement. At the same time, we find that labels applied with confident ratings tend to be more accurate (table 1). Thus we note an interesting pattern in our crowdsourced data: the average confidence rating of a sentence is an indirect but rather effective estimate of the accuracy of the sentence's label. This suggests that we can increase the accuracy of our training data by filtering out worker annotations that are given with low confidence.

Confidence Vote	Accuracy	
	In-Class	Threshold
Very Confident	0.975	0.975
Somewhat Confident	0.864	0.953
Not Confident	0.534	0.941

Table 1: Crowd accuracy by confidence rating. 'In-class' accuracy gives the percent of crowd labels with that exact confidence rating that matched the gold standard label; 'threshold' refers to the percent of labels with the same or more confident rating that matched the gold standard.

3.3 Using annotations to train a classifier

To see whether the crowdsourced dataset is as useful for training a classification model as the expert-labeled dataset, we conducted three experiments:

3.3.1 Experts vs The Crowd

First, we train two versions of our classification model: one using only gold standard labels as training data (expert-trained classifier), and the other using only crowdsourced labels (crowd-trained classifier). Each classifier uses the same number of training instances.

Since the gold standard data set is so small, we use stratified K-fold cross validation (k=5) to train the expert-trained classifier on different portions of the gold standard data set (Hastie et al., 2009). For each K-fold iteration, we also randomly sub-sample (with replacement) a training set from the crowdsourced data of equal size (~260 samples), and evaluate both classifiers against the validation portion of the gold standard data.

As detailed in table 2, the average accuracy of the expert-trained classifier is 0.84 ($\pm$.04), and the average accuracy of the crowd-trained classifier is 0.86 ($\pm$.03). There is no significant difference between these two classifiers, which shows that the crowdsourced dataset is just as useful for training a classification model as the expert-labeled dataset given the same number of training instances.

3.3.2 Increasing training instances

To test whether we can improve the accuracy of the classification model by simply increasing the number of crowd worker annotations we collect, we train classifiers using increasing training set sizes. For each size we randomly sub-sample a training set from the crowdsourced labels and

Training Set	Size	Accuracy	F-Score
Gold	259	$0.84 \pm .04$	$0.84 \pm .03$
Crowd	259	$0.86 \pm .03$	$0.87 \pm .04$

Table 2: Experts vs The Crowd Results

evaluate it against the entire gold standard dataset. Figure 3 below shows the mean and 90% confidence interval for accuracy over 50 random subsamples at each training set size. Performance improves with the size of the training set, but begins to level off when we use all available crowdsourced labels (training set size 717). This suggests that we might achieve only modest improvements in accuracy by gathering further crowdsourced labels.

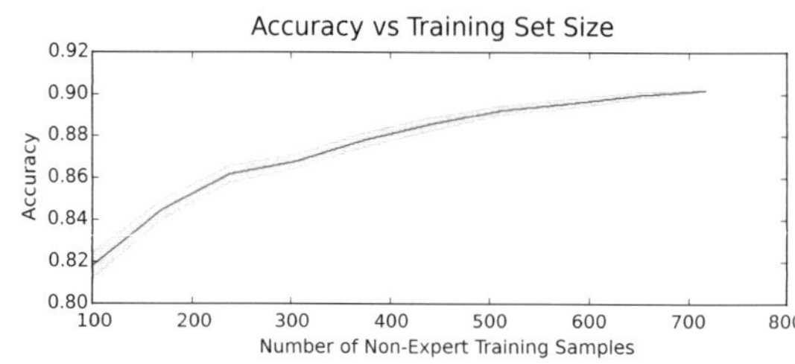

Figure 3: Classifier accuracy by training size. Performance improves with the size of the training set, but begins to level off around 700 samples.

3.3.3 Incorporating confidence thresholds

We observed that crowd annotations with *Very Confident* votes tend to be more accurate than those with less confident votes when evaluated against a gold standard (table 1). Our third experiment tests whether limiting the crowdsourced training data to incorporate only worker labels given with high confidence will improve the classifier's accuracy.

We train our model on three further training sets with increasing confidence thresholds. When evaluated against the entire gold standard test set, the classifier trained under the *Not Confident* threshold, which includes all of the training sentences, achieves an accuracy of 0.90. The classifier trained under the *Somewhat Confident* threshold receives a modest boost in accuracy (0.91), even though there are fewer training samples available at that threshold. While the *Very Confident* threshold classifier achieves the highest precision (figure 4), its high threshold limits the number of training instances available and thus produces a lower accuracy and F-Score. (In fact, if we restrict the number of training samples under each threshold to 532, the accuracy of the *Not Confident* and *Somewhat Confident* thresholds drop to 0.86 and 0.85 respectively.) Overall, the *Somewhat Confident* training set, which balances training set size and label confidence, produced the optimal outcome.

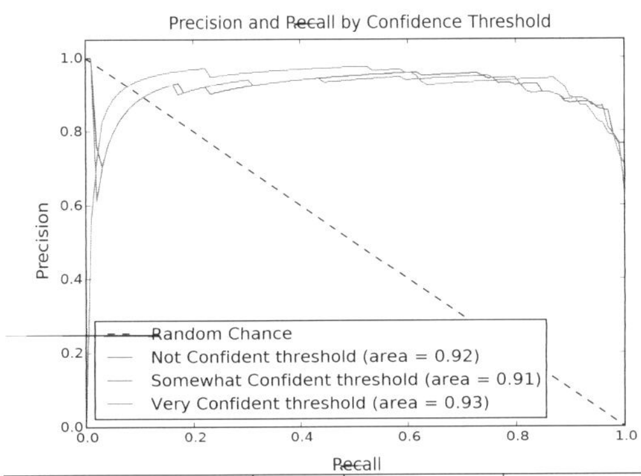

Training Thresh.	Size	Accuracy	F-Score
Not Conf.	717	0.90	0.91
Somewhat Conf.	690	0.91	0.91
Very Conf.	532	0.88	0.89

Figure 4: Training Set Confidence Thresholds

4 Discussion

A limitation of this study is that some sentences in AudGenDB are readily classifiable by non-experts due to their lexical content or syntactic structure. Though this requires further research, we conducted a preliminary analysis to explore the impact this may have had on our results. Running the Stanford CoreNLP pre-trained sentiment prediction model (Socher et al., 2013) on our gold dataset and assigning a *normal* label to sentences predicted *positive* or *neutral* by CoreNLP and an *abnormal* label to sentences predicted *negative* produces output that is 70.4% accurate[2]. But if we use the average time spent by workers in classifying each sentence as a rough indicator of difficulty, we see that 'easier' sentences (those taking less than 60 seconds to classify on average) are more accurately labeled by the sentiment analysis model than more 'difficult' sentences (76.4% vs 69.5% accuracy respectively). Thus, it appears that the hardest sentences to classify are less clearly *normal* or *abnormal* based on lexical content or syntactic structure alone.

Our results show that it is possible to use crowdsourcing to generate sentence labels for a task that requires specific domain knowledge. By applying labels to sentences based on a weighted vote of the crowd annotators, we can generate a training dataset that is as effective as one generated by expert annotators in training a sentence classifier. We can improve the usefulness of the crowdsourced dataset by simply gathering additional annotations, to a point. When gains in accuracy achieved through growing the training set begin to level off, we can improve classifier accuracy further – without collecting more data – by incorporating individual crowd confidence ratings.

5 Related Work

There has been considerable research effort aimed at reducing the infamously high monetary and time cost of expert data annotation. Some studies examine ways to optimize accuracy of expert annotations with minimal cost (Grouin et al., 2014; Rzhetsky et al., 2009)]. Other research, such as this work, focuses on crowdsourcing as a way of reducing annotation cost.

Crowdsourcing is rapidly growing as a data collection method in bioinformatics (Khare et al., 2015). Within the biomedical crowdsourcing literature, methods for outsourcing tasks that require domain knowledge generally fall into one of two categories. The first type uses active crowdsourcing platforms to locate domain experts within the crowd (Ipeirotis and Gabrilovich , 2014; Shapiro et al., 2013; CrowdMed, 2015). The second focuses on harnessing the efforts of non-experts in various ways. Some researchers have leveled the playing field between experts and crowdsourced humans by gamifying complex tasks (Cooper et al., 2010) or simply training crowdsourced workers to complete tasks with limited scope (Warby et al., 2014). In some cases, crowdsourced humans turn out to be just as accurate on their own as experts (Zhai et al., 2013). In others, researchers aggregate crowdsourced annotations to produce a dataset that approaches the accuracy of an expert-generated gold standard (MacLean and Heer, 2013). This work falls firmly into this last group.

[2]Before running sentiment prediction, we replaced words that are uniquely positive in our dataset like *unremarkable* and *patent* with equivalent words like *good* that are more commonly positive in the online reviews on which the model was trained. See supplemental material for details.

References

Charu C. Aggarwal and ChengXiang Zhai. 2012. A Survey of Text Classification Algorithms. *Mining Text Data*, 163–122. Springer Science & Business Media, New York, NY.

The Audiological and Genetic Database (AudGenDB) [Internet]. Philadelphia: The Children's Hospital of Philadelphia. 2006 – [cited 2015 Aug 10]. Available from: http://http://audgendb.chop.edu/.

Aris Anagnostopoulos, Luca Becchetti, Adriano Fazzone, Ida Mele, and Matteo Riondato. 2015. The Importance of Being Expert: Efficient Max-Finding in Crowdsourcing. In *Proceedings of the 2015 ACM SIGMOD Itnernational Conference on Management of Data*, 983–998.

Seth Cooper, Firas Khatib, Adrien Treuille, Janos Barbero, Jeehyung Lee, Michael Beenen, Andrew Leaver-Fay, David Baker, Zoran Popovi, and Foldit Players. 2010. Predicting protein structures with a multiplayer online game. *Nature*, 466(7307): 756–760.

CrowdMed. *CrowdMed*. http://www.crowdmed.com. Web. 29 July 2015.

Cyril Grouin, Thomas Lavergne, and Aurélie Névéol. 2014. Optimizing annotation efforts to build reliable annotated corpora for training statistical models. In *LAW VIII–The 8th Linguistic Annotation Workshop*, 2014: 54–58.

Trevor Hastie, Robert Tibshirani, and Jerome Friedman. 2009. "Cross-Validation". In *Elements of Statistical Learning, 2nd Ed.* (241–249). New York, NY, USA: Springer New York Inc.

Pei-Yun Hsueh, Prem Melville, and Vikas Sindhwani. 2009. Data quality from crowdsourcing: a study of annotation selection criteria. In *Proceedings of the NAACL HLT 2009 Workshop on Active Learning for Natural Language Processing*, 27–35.

Panagiotis G. Ipeirotis and Evgeniy Gabrilovich. 2014. Quizz: Targeted Crowdsourcing with a Billion (Potential) Users. In *Proceedings of WWW'14 23rd International World Wide Web Conference*, 143–154.

Adam Kapelner, Krishna Kaliannan, Dean Foster, and Lyle Ungar. 2012. When is Word Sense Disambiguation Difficult? A Crowdsourcing Approach. (Working paper) *Wharton Research Scholars Journal*, 6-26-2012. http://repository.upenn.edu/wharton_research_scholars/116.

Ritu Khare, Benjamin M. Good, Robert Leaman, Andrew I. Su, and Zhiyong Lu. 2015. Crowdsourcing in biomedicine: challenges and opportunities. *Briefings in Bioinformatics*, 2015: 1–10.

Richard J. Landis and Gary G. Koch. 1977. The Measurement of Observer Agreement for Categorical Data. *Biometrics*, 33(1): 159–174.

Diana Lynn MacLean and Jeffrey Heer. 2013. Identifying medical terms in patient-authored text: a crowdsourcing-based approach. *Journal of the American Medical Informatics Association*, 20(6): 1120–1127.

Christopher D. Manning, Mihai Surdeanu, John Bauer, Jenny Finkel, Steven J. Bethard, and David McClosky. 2014. The Stanford CoreNLP Natural Language Processing Toolkit. In *Proceedings of 52nd Annual Meeting of the Association for Computational Linguistics: System Demonstrations*, 55–60.

Andrey Rzhetsky, Hagit Shatkay, and W. John Wilbur. 2009. How to Get the Most out of Your Curation Effort. *PLoS Computational Biology*, 5(5): e1000391.

Danielle N. Shapiro, Jesse Chandler, and Pam A. Mueller. 2013. Using Mechanical Turk to Study Clinical Populations. *Clinical Psychological Science*, 1(2): 213–220.

Richard Socher, Alex Perelygin, Jean Wu, Jason Chuang, Christopher Manning, Andrew Ng, and Christopher Potts. 2013. Recursive Deep Models for Semantic Compositionality Over a Sentiment Treebank. In *Proceedings of the 2013 Conference on Empirical Methods in Natural Language Processing*, 254–263

Rion Snow, Brendan O'Connor, Daniel Jurafsky, and Andrew Y. Ng. 2008. Cheap and Fast–But is it Good? Evaluating Non-Expert Annotations for Natural Language Tasks. In *Proceedings of the 2008 Conference on Empirical Methods in Natural Language Processing*, 254–263.

Aobo Wang, Cong Duy Vu Hoang, and Min-Yen Kan. 2013. Perspectives on crowdsourcing annotations for natural language processing. *Language Resources and Evaluation*, 47(1): 9–31.

Simon C. Warby, Sabrina L. Wendt, Peter Welinder, Emil G.S. Munk, Oscar Carrillo, Helge B.D. Sorensen, Poul Jennum, Paul E. Peppard, Pietro Perona, and Emmanuel Mignot. 2014. Sleep spindle detection: crowdsourcing and evaluating performance of experts, non-experts, and automated methods. *Nature Methods*, 11(4): 385–392.

Haijun Zhai, Todd Lingren, Louise Deleger, Qi Li, Megan Kaiser, Laura Stoutenborough, and Imre Solti. 2013. Web 2.0-Based Crowdsourcing for High-Quality Gold Standard Development in Clinical Natural Language Processing. *Journal of Medical Internet Research*, 15(4): e73.

Identifying Key Concepts from EHR Notes Using Domain Adaptation

Jiaping Zheng
College of Information
and Computer Sciences
University of Massachusetts
Amherst, MA
jzheng@cs.umass.edu

Hong Yu
Bedford VA Medical Center
Bedford, MA
Department of Quantitative Health Sciences
University of Massachusetts
Worcester, MA
hong.yu@umassmed.edu

Abstract

Linking electronic health records (EHRs) to relevant education materials can provide patient-centered tailored education which can potentially improve patients' medical knowledge, self-management and clinical outcome. It is shown that EHR query generation using key concept identification improves retrieval of education materials. In this study, we explored domain adaptation approaches to improve key concept identification. Our experiments show that a 20.7% improvement in the F1 measure can be achieved by leveraging data from Wikipedia. Queries generated from the best performing approach achieved a 20.6% and 27.8% improvement over the queries generated from the baseline approach.

1 Introduction

Providing patients with access to their own electronic health records (EHRs) has been shown to benefit patients in many ways, including enhanced medical understanding, and better medication adherence (Delbanco et al., 2012). Several studies have also found that providing knowledge can improve diabetes-related health outcomes (Wiljer et al., 2006).

However, EHR notes present unique challenges to the average patients. A national survey in US shows that 36% of the population have basic or below basic health literacy (National Center for Education Statistics, 2003). The language in the EHR notes is difficult for non-medical professionals to comprehend because of the prevalence of medical terms, abbreviations, and domain-specific language patterns. Coupled with limited average health literacy, the valuable and authoritative information contained in the EHR is less accessible to the patients, who ultimately stand to benefit the most from the information.

Linking EHR notes to relevant education materials can unlock the information in them and provide patient-centered tailored education which has the potential to enhance patient engagement and lead to improved self-management and clinical outcomes. One challenge in designing such an Information Retrieval system is to generate queries. It is shown that ad hoc retrieval using the entire EHR note is less effective because of the noise contained in the notes (Zheng and Yu, 2015). A better strategy is to identify the key concepts from the notes and use them as queries. Using off-the-shelf concept recognition tools such as MetaMap (Aronson, 2001) can lead to long queries that contain many unimportant concepts. For example, incidental findings in an EHR note may distract the retrieval system from returning documents that are central to the note. Therefore, identifying, among all the concepts, the important ones is essential to generate effective queries. In this study, we explored domain adaptation approaches (Jiang and Zhai, 2007; Daumé III, 2007) to improve key concept identification. These approaches have been demonstrated to improve performances of NLP tasks such as semantic role labeling (Dahlmeier and Ng, 2010) and discourse connective detection (Polepalli Ramesh et al., 2012).

Our system in the training phase uses a combination of Wikipedia data and EHR data to learn models to identify key concepts. At the test time, the models are used to predict key concepts from the EHR notes. The identified key concepts are then grouped into one query string to retrieve relevant education documents.

2 Related Work

Domain adaptation is a method to adapt machine learning models trained from a large labeled out-

Proceedings of the Sixth International Workshop on Health Text Mining and Information Analysis (Louhi), pages 115–119,
Lisbon, Portugal, 17 September 2015.

of-domain dataset to a target domain in which labeled data is difficult to obtain. Due to privacy regulations, limited health care data is readily available to train machine learning models (Chapman et al., 2011). Thus, domain adaptation approaches are investigated in many NLP tasks. In Part of Speech tagging, Coden et al. (2005) showed combining Penn Treebank data with a small clinical notes corpus improves performance. Liu et al. (2007) developed a heuristic sample selection method to select training samples from the medical domain, and combined with Penn Treebank data to adapt a maximum entropy tagger.

There is also interest in adapting models in other NLP tasks. Polepalli Ramesh et al. (2012) showed that domain adaptation techniques yielded the best performance in identifying discourse connectives in biomedical text. Kim et al. (2013) extracted congestive heart failure related mentions by adapting models learned from a different type of clinical notes.

Information Retrieval in the biomedical domain is also related to this work. The CLEF eHealth (Kelly et al., 2014) challenge includes a task to retrieve information to address questions patients may have when reading clinical reports. This task provides participants with expert-formulated concise queries for one central disorder in discharge summaries (Goeuriot et al., 2014). In our study, we aim to generate queries from long EHR notes without the help of experts. TREC Clinical Decision Support Track is another information retrieval challenge involving EHR notes. The task is designed to address the physicians' information needs rather than the patients'. Case reports are provided as query descriptions, which can be shorter and more focused than an EHR note.

3 Materials

Twenty progress notes are randomly selected from a de-identified corpus of EHR notes to test our systems' performance. Each note contains on average 261 tokens, with a standard deviation of 133. A physician read each note, and manually identified relevant education materials from a collection of MedlinePlus[1] documents. The phrases in the EHR notes that match the title of a relevant MedlinePlus document are marked as key concepts. A snippet of one note with its linked education materials is show in Table 1. Key concepts marked by matching titles are *italicized*.

Snippet of EHR
Patient is a XX-year-old woman status post Thoratec left ventricular assist device placement for *cardiogenic shock* following acute myocardial infarction. Patient requires critical care for management of her *respiratory failure*, malnutrition, hyperglycemia, post-procedure hemodynamics, and *renal failure*.
Select Relevant Education Materials
Heart Attack Cardiogenic shock Kidney Failure Respiratory Failure

Table 1: Snippet of an EHR note and titles of its linked MedlinePlus documents.

For domain adaptation, we collected Wikipedia articles that are in the Diabetes category. This ensures the Wikipedia articles are from the same domain. The internal Wikipedia links in each article are used as key concepts. There are a total of 130 Wikipedia articles.

The education material collection to evaluate retrieval performance consists of approximately 9400 documents from the "Health Topics", "Drugs and Supplements", and "Medical Encyclopedia" sections of the MedlinePlus website. On average, the documents have 749 tokens, with a standard deviation of 566.

4 Methods

4.1 Domain Adaptation Approaches

We trained Conditional Random Fields (CRF) models to predict the key concepts. As a baseline system, we used leave-one-out cross validation on the EHR notes. The features in the model include lexical, capitalization, prefix, suffix, word shape, and UMLS semantic type. The semantic types are provided by MetaMap, and added as a feature to each token of the MetaMap-recognized terms.

We compared three different methods of domain adaptation to identify the key concepts—instance weighting, instance pruning, and feature augmentation. In accordance with the common terminology, we refer to the larger Wikipedia data as source domain, and the smaller 20 EHR notes the target domain data.

[1]http://www.nlm.nih.gov/medlineplus/

Instance weighting (Jiang and Zhai, 2007) merges the data from both corpora with different weights during training. The weights are usually inversely proportional to the size of the corpus. A model is then trained using this weighted training dataset. In our experiments, we used leave-one-out cross validation on the target domain data. In each fold, the training data is a weighted combination of the Wikipedia data and 19 EHR notes. The test data is the left out EHR note.

Instance pruning (Jiang and Zhai, 2007) removes misleading training instances from the source domain by first applying a model learned from the target domain. For example, if an instance is assigned different labels in the source and target domain corpora, it is removed to prevent the algorithm from learning from this confusing data. We first trained a model on the target domain data, and then predicted the labels on the source domain data. Instances in the source domain that were incorrectly labeled were pruned from the source training set. Finally, a new model was trained using this pruned source domain dataset.

Feature augmentation (Daumé III, 2007) adds additional features to the training instances to identify which corpus they come from. For each original feature in a training example, a new indicator feature is included to indicate the origin domain of the feature, so the learning algorithm can distinguish features important to each domain. A model is then trained on the combined dataset. In our experiments, we applied cross validation on the target domain in a similar fashion to the instance weighting experiments. In each fold, a feature-augmented corpus was built from all the Wikipedia data and 19 EHR notes, and the test data consisted of one EHR note.

4.2 Query Generation

To evaluate the key concepts' effectiveness on education material retrieval, we used the key concepts as queries. The textual MedlinePlus documents are indexed using Galago (Croft et al., 2010), an open source system. In the instance weighting and feature augmentation experiments, the predicted key concepts in the left out EHR note in each fold are combined as queries. In the instance pruning experiments, the predicted key concepts in the EHR notes using the pruned source domain data are used as queries.

Following the same design as reported in Zheng and Yu (2015), we experimented with a two-stage approach, using the same parameters. This approach first issues a query using the key concepts, and then issues a second query using all the concepts recognized by MetaMap. The top 20 results from the first query and the results from the second query are merged to be the final result, removing duplicates between the two result sets.

In all the IR systems, we use Mean Average Precision (MAP) (Manning et al., 2008), a common metric in the IR community, to evaluate the ranked retrieval results. Set-based measures such as precision and recall metrics cannot distinguish the order the results are presented in a ranked retrieval context.

System	Precision	Recall	F1
Baseline	45.77%	26.51%	31.76
Instance Weighting	47.59%	34.41%	38.32
Instance Pruning	40.00%	6.02%	10.23
Feature Augmentation	46.60%	28.86%	34.08

Table 2: Key concept identification results.

5 Results

The results of the baseline CRF model and the models using domain adaptation approaches are shown in Table 2. The baseline system achieved an F1 score of 31.76. Two domain adaptation approaches, instance weighting and feature augmentation, outperformed the baseline system. Both the precision and recall were improved in these two approaches. The best performing approach (instance weighting) shows a 6.56 points (20.7%) improvement in F1 measure over the baseline system.

The Information Retrieval results using these key concepts as queries are shown in the "MAP" column in Table 3. Queries generated from the instance weighting approach outperformed the baseline query results by 0.019 points (20.6%). The other two approaches did not improve over the baseline query.

Results using two-stage approach is shown in the "Two Stage" column of Table 3. The instance weighting approach again outperformed the baseline approach by 0.031 points (27.8%). The other two approaches' performances were similar to the baseline result.

Table 3: Information Retrieval performance. In each system, the queries are generated by combining the recognized key concepts.

Queries	MAP	Two Stage
Baseline	0.0921	0.1114
Instance Weighting	0.1111	0.1424
Instance Pruning	0.0316	0.1002
Feature Augmentation	0.0684	0.1081

6 Discussions

In the domain adaptation experiments, the precision of the three approaches were relatively close to the baseline. However, the recall scores vary greatly. In the instance weighting experiment, the model was able to identify many abbreviations that are rare in the target domain. For example, "EGD" and "DVT" were successfully identified as key concepts despite their occurring only once and three times in the target domain corpus. On the other hand, the instance pruning approach removed over half of the training instances from the source domain data, resulting in a lower performance. The Wikipedia Manual of Style states that only the first occurrence of a term should be linked, and generally a link should only appear once. This resulted in many valid instances being removed because of multiple occurrences. For example, repeated mentions of "glucose" in Wikipedia articles were predicted as key concepts by the target domain model. However, most were removed because only one of them in each article was linked to the glucose article. The reduced training size lowered the recall of this model.

In the IR experiments, the instance weighting approach outperformed the baseline in both the single query and the two stage designs. This can be attributed to the higher recall of this approach in the CRF model. Due to its low recall in key concept identification, instance pruning failed to retrieve many relevant documents. For example, in six of the EHR notes, only one phrase was labeled as key concept, and one of them was incorrect. Despite feature augmentation's improvement in the key concept identification experiments over the baseline, queries generated from this approach did not improve over the baseline query result. The identified key concepts by this method included abbreviations such as "CHF" and general symptoms such as "nausea", which can be associated with a multitude of diseases.

One limitation of the study is that the retrieval gold standard was annotated by one physician. Additional annotators would produce better annotations.

7 Conclusion

It is shown that identifying the key concepts is an effective strategy to generate queries to link EHR notes to education materials. In this study, we explored several domain adaptation approaches to improve key concept identification from EHR notes. The source domain data from Wikipedia enabled the CRF models to learn from more examples. Our experiments have shown that the best setup outperformed a baseline CRF system by 20.7% using data from Wikipedia. Using key concepts recognized from this setup resulted in the best information retrieval performance, a 20.6% improvement over the baseline. Under a two-stage query strategy, retrieval results using these key concepts outperformed the baseline by 27.8%.

Acknowledgments

This work was supported in part by the Award 1I01HX001457 from the United States Department of Veterans Affairs Health Services Research and Development) Program Investigator Initiated Research. The contents do not represent the views of the U.S. Department of Veterans Affairs or the United States Government.

References

Alan R. Aronson. 2001. Effective mapping of biomedical text to the UMLS Metathesaurus: the MetaMap program. *Proc AMIA Symp*, pages 17–21.

Wendy W. Chapman, Prakash M. Nadkarni, Lynette Hirschman, Leonard W. D'Avolio, Guergana K. Savova, and Ozlem Uzuner. 2011. Overcoming barriers to NLP for clinical text: the role of shared tasks and the need for additional creative solutions. *Journal of the American Medical Informatics Association*, 18(5):540–543.

Anni R. Coden, Serguei V. Pakhomov, Rie K. Ando, Patrick H. Duffy, and Christopher G. Chute. 2005. Domain-specific language models and lexicons for tagging. *Journal of Biomedical Informatics*, 38(6):422–430.

W. Bruce Croft, Donald Metzler, and Trevor Strohman. 2010. *Search Engines: Information Retrieval in Practice*. Addison-Wesley.

Daniel Dahlmeier and Hwee Tou Ng. 2010. Domain adaptation for semantic role labeling in the biomedical domain. *Bioinformatics*, 26(8):1098–1104.

Hal Daumé III. 2007. Frustratingly easy domain adaptation. In *Proceedings of the 45th Annual Meeting of the Association of Computational Linguistics*, pages 256–263, June.

Tom Delbanco, Jan Walker, Sigall K. Bell, Jonathan D. Darer, Joann G. Elmore, Nadine Farag, Henry J. Feldman, Roanne Mejilla, Long Ngo, James D. Ralston, Stephen E. Ross, Neha Trivedi, Elisabeth Vodicka, and Suzanne G. Leveille. 2012. Inviting patients to read their doctors' notes: a quasi-experimental study and a look ahead. *Ann. Intern. Med.*, 157(7):461–470, October.

Lorraine Goeuriot, Liadh Kelly, Wei Li, Jo ao Palotti, Pavel Pecina, Guido Zuccon, Allan Hanbury, Gareth Jones, and Henning Müller. 2014. ShARe/CLEF eHealth evaluation lab 2014, task 3: User-centred health Information Retrieval. In *CEUR Workshop Proceedings*, volume 1180, pages 43–61.

Jing Jiang and ChengXiang Zhai. 2007. Instance weighting for domain adaptation in NLP. In *Proceedings of the 45th Annual Meeting of the Association of Computational Linguistics*, pages 264–271, June.

Liadh Kelly, Lorraine Goeuriot, Hanna Suominen, Tobias Schreck, Gondy Leroy, DanielleL. Mowery, Sumithra Velupillai, Wendy W Chapman, David Martinez, Guido Zuccon, and João Palotti. 2014. Overview of the ShARe/CLEF eHealth evaluation lab 2014. In Evangelos Kanoulas, Mihai Lupu, Paul Clough, Mark Sanderson, Mark Hall, Allan Hanbury, and Elaine Toms, editors, *Information Access Evaluation. Multilinguality, Multimodality, and Interaction*, volume 8685 of *Lecture Notes in Computer Science*, pages 172–191. Springer International Publishing.

Youngjun Kim, Jennifer Garvin, Julia. Heavirland, and Stéphane M. Meystre. 2013. Improving heart failure information extraction by domain adaptation. *Stud Health Technol Inform*, 192:185–189.

Kaihong Liu, Wendy Chapman, Rebecca Hwa, and Rebecca S. Crowley. 2007. Heuristic sample selection to minimize reference standard training set for a part-of-speech tagger. *Journal of the American Medical Informatics Association*, 14(5):641–650.

Christopher D. Manning, Prabhakar Raghavan, and Hinrich Schütze. 2008. *Introduction to Information Retrieval*. Cambridge University Press.

National Center for Education Statistics. 2003. National assessment of adult literacy (NAAL).

Balaji Polepalli Ramesh, Rashmi Prasad, Tim Miller, Brian Harrington, and Hong Yu. 2012. Automatic discourse connective detection in biomedical text. *Journal of the American Medical Informatics Association*, 19(5):800–808.

David Wiljer, Sima Bogomilsky, Pamela Catton, Cindy Murray, Janice Stewart, and Mark Minden. 2006. Getting results for hematology patients through access to the electronic health record. *Can Oncol Nurs J*, 16(3):154–164.

Jiaping Zheng and Hong Yu. 2015. Methods for linking EHR notes to education materials. In *Proc. AMIA Summit on Clinical Research Informatics*, pages 209 – 215.

Information Extraction from Biomedical Texts: Learning Models with Limited Supervision

Marie-Francine Moens
KULeuven, Leuven, Belgium
sien.moens@cs.kuleuven.be

Abstract

Among the application domains of information extraction, the biomedical domain is one of the most important ones. This is due to the large amount of biomedical text sources including the vast scientific literature and collections of patient reports written in natural language. These sources contain a wealth of crucial knowledge that needs to be mined. Typical mining tasks regard entity recognition, entity-relation extraction, and event and event participant recognition. Recently we witness an interest in the recognition of spatial relationships between entities and of temporal relationships between events. One of the most important problems in information extraction regards dealing with a limited amount of examples that are manually annotated by experts and that can be used for training the extraction models.

In this talk we discuss how we can leverage knowledge contained in unlabelled texts and ontological knowledge about known relationships between the output labels used for the extractions. The former aspect especially focuses on how to automatically create novel training examples from the unlabelled data, the latter on how to integrate the relationships in models for structured machine learning during training and testing of the extraction models in the most efficient way. We show promising results and point to directions of future research.

Proceedings of the Sixth International Workshop on Health Text Mining and Information Analysis (Louhi), page 120,
Lisbon, Portugal, 17 September 2015.

*** Formatting Issue - Best Available Paper/Graphic ***

Adverse drug event classification of health records using dictionary-based pre-processing and machine learning

Stefanie Friedrich and Hercules Dalianis
Department of Computer and Systems Sciences (DSV)
Stockholm University
P.O. Box 7003
164 07 Kista
Sweden
stfr6041@student.su.se, hercules@dsv.su.se

Abstract

A method to find adverse drug reactions in electronic health records written in Swedish is presented. A total of 14,751 health records were manually classified into four groups. The records are normalised by pre-processing using both dictionaries and manually created word lists. Three different supervised machine learning algorithm were used to find the best results; decision tree, random forest and LibSVM. The best performance on a test dataset was with LibSVM obtaining a precision of 0.69 and a recall of 0.66, and a F-score of 0.67. Our method found 865 of 981 true positives (88.2%) in a 3-class dataset which is an improvement of 49.5% over previous approaches.

1 Introduction

Adverse drug events (ADEs) are the seventh largest cause of death in Sweden (Wester et al., 2008). ADEs also cause 3.7% of hospital admissions worldwide (Howard et al., 2007). Drugs have been developed by pharmaceutical companies and tested on a small group of healthy young men (~3,000) (FDA, 2014); however, patients taking the drugs are mostly elderly and multiple sick people, therefore, one needs to perform post-marketing drug safety surveillance in order to detect the ADEs' effect on real patients.

The care of the patient is continuously documented by the physician in health records also called electronic patient records. The health records contain both structured and unstructured information. The structured information is for example the age of the patient, time stamps, drugs, ICD-10 diagnosis code, and microbiological and blood values. Unstructured information is mainly free text. Health records are usually long and written by different authors with different writing styles (Allvin et al., 2011; Wijesekera, 2013). To identify entities in a text, to extract meaning and terms, and to consider their context, advanced methods must carried out. Several strategies and methods have been developed, and some approaches are described in the next section.

2 Related research

There are several studies on automatically identifying ADEs from the text of electronic health records using either rule-based or machine learning-based methods. In this section different approaches and their results are summarised.

2.1 Rule based methods

Several rule-based studies to detect ADEs have been carried out. Eriksson et al. (2013), carried out a rule and dictionary approach to detect ADEs in 6,011 Danish psychiatric patients' hospital records. The system identified 35,477 unique ADEs. They obtained a precision of 0.89 and a recall of 0.75.

Wang et al. (2009) developed a rule-based system to detect the *drug - ADEs relationship* for seven specific drugs. They used 25,074 discharge summaries in English to evaluate the system. The authors obtained a recall and precision of 0.75 and 0.31, respectively, for known ADEs.

Hazlehurst et al. (2009) detected vaccine ADEs among Kaiser Permanente Northwest (KPNW), which encompasses more than 450,000 persons. They compared automated methods MediClass with code-based detection methods; the MediClass method obtained better results than the code-based method - 0.74 versus 0.31 PPV (positive predictive value, which is the same as precision).

Proceedings of the Sixth International Workshop on Health Text Mining and Information Analysis (Louhi), pages 121–130, Lisbon, Portugal, 17 September 2015.

2.2 Machine learning based methods

There are several studies on automatically identifying ADEs from text of electronic patient records. One Spanish study by Santiso et al. (2014) used 6,100 concepts and 4,700 adverse drug reactions (ADRs) relations for training, and evaluated on 2,100 concepts and 1,600 ADR relations, and obtained 0.93 precision and 0.85 recall using the Random Forest algorithm.

In a Japanese study by Aramaki et al. (2010), using 3,012 Japanese discharge summaries, the authors annotated 1,045 drugs and 3,601 possible adverse drug effects. They found that around 7.7% of the discharge summaries contained ADE. Of these, 59% could be extracted automatically. They used both support vector machine (SVM) and pattern matching methods(PTM) and obtained slightly better results using PTM; precision was 0.41 and recall 0.92 when using PTM, and SVM gave precision of 0.58 and recall of 0.62.

In a study by Roller and Stevenson (2014), UMLS was used to identify concepts and relations in millions of biomedical articles (for instance, drug contraindications, ADE drug relations), and used them to train a Naïve Bayes classifier obtaining 0.25 precision and 1.00 recall.

Gurulingappa et al. (2012) developed a manually annotated corpora in English consisting of 3,000 medical case reports (i.e. published scientific reports of specific patients, their drugs and their side effects). Three annotators annotated the corpora using the concepts *drugs, drug dosage, adverse effect* and *relationship* among the concepts. The three annotators have an M.Sc. degree in biomedicine, where two of them were experienced and one novice. One annotator was used as standard and the other as reference. Each annotator annotated 2,000 documents, and 1,000 documents were annotated by all three annotators. The F-score was measured. *Drugs* obtained an F-score for partial match from 0.90 down to 0.38. *Adverse effect-Drugs* obtained an F-score of 0.79 down to 0.37. The ADE-corpus is publicly available[1]. The authors performed machine learning experiments both with Naïve-Bayes and Maximum Entropy (MaxEnt) classifiers from the MALLET toolkit, and obtained, as best for MaxEnt, 0.75 precision and 0.64 recall.

[1]https://sites.google.com/site/adecorpus/

2.3 Aim and purpose

Previous approaches to detect adverse events have used either rule-based approaches or machine learning approaches, but none have applied a mixed method.

We aimed to design a method that identifies ADRs in health records. The identification of ADRs is realised with a mixture of keyword and phrase filtering and supervised machine learning algorithms that classify health records. By filtering of ADR related phrases, we achieved less computational effort to obtain the prediction and higher prediction performance. We also aspired to design a flexible method that is able to distinguish between different kinds of ADRs - for example, possible ADRs and ADRs related to a certain drug. Finally, we strived for both classification of medical records concerning ADRs and revealing the drug-symptom relations that are decisive for this classification.

3 Materials and methods

3.1 SEPR Corpus and SEPR Drug Corpus

Stockholm Electronic Patient Record (SEPR) Corpus is a patient record collection encompassing over one million patient records from the years 2006-2014 from Karolinska University Hospital in Stockholm, (Dalianis et al., 2012). Of this SEPR Corpus[2], records were sampled to be used for the machine learning experiment.

The SEPR Corpus is stored in a relational database. The unique serial number of each patient was extracted to identify the corresponding health record written by the physicians. Each entry of the record was ordered in temporal order including the drugs taken by each patient. Although, there are no personal names in the data base there can be personal names mentioned in the free text and therefore the data still may contain confidential information. Thus, the data cannot be made publicly available. The problem is known and there are initiatives to establish an infrastructure of publicly available medical records for research (Dalianis et al., 2015).

[2]This research has been approved by the Regional Ethical Review Board in Stockholm, (Etikprövningsnämnden i Stockholm), permission number 2012/834-31/5.

*** Formatting Issue - Best Available Paper/Graphic ***

3. **Oxynorm** (ATC code N02AA05), analgetica (Swedish: smärtstillande medel), also called a pain killer; drug agent Oxycodone
4. **Prednisolon** (ATC code H02AB06), corticosteroids (Swedish: Kortikosteroider för systemiskt bruk), also called cortisone; drug agent Prednisolon
5. **Primperan** (ATC code A03FA01), functional gastrointestinal disorders (Swedish: funktionella mag-tarmsymptom); to treat nausea and vomiting, drug agent: Metoclopramide

Dosage and dosage form were not considered, nor the companies producing or dealing with any of the above mentioned drugs.

As a second sampling step, four classes were defined, and assigned to health records. Health records that could not be assigned to one of the four classes were excluded; these were mainly health records with contradicting notes about ADRs. The four classes are:

1. **definitely ADRs and related to the chosen drugs**: known and drug-related ADRs according to FASS (2014b)
2. **definitely ADRs but not related to the chosen drug**: ADR because an ADR term (in Swedish e.g. biverkning, läkemedelutlöst) is mentioned in the health record, or diagnosis related to ADR (e.g. IDC G44.4 Drug-induced headache, not elsewhere classified); category A1 according to Stausberg and Hasford (2011)
3. **possible ADRs**: suspicious reaction or suspected adverse drug event, however, neither diagnosis is related to ADR or symptoms are linked to ADR
4. **no ADR**: 'clean' patients with no ADR mentioned, no ADR related diagnosis, or ADR related symptom

3.2.2 Manual class labelling

The 14,751 health records from the SEPR Drug Corpus were manually classified by one annotator who also is a computer scientist. Large parts of the records do not contain any note about ADRs. To aid the annotator the records were pre-annotated; with pre-annotation, only ADR-related passages have to be read. Designed word lists were used, which supported the pre-annotation and the manually performed class labelling. The manual annotation took around seven weeks to perform.

With the performed two sampling steps (five drugs, four classes) the SEPR Drug Corpus was constructed containing 14,751 health records assigned to four class labels, presented in Table 1.

Stausberg and Hasford (2011) categorised 505 ICD-10 codes in seven groups "with respect to its validity as an indicator for an ADE and its definition in the ICD-10". To avoid biased class labelling, patients with a diagnosis related to Stausberg's category A.1 (drug-related causation was noted in the ICD-10) were assigned to class 2 regardless of whether ADRs were mentioned in the health record.

3.3 Text manipulation

To improve both prediction performance and computational speed, normalisation and pre-processing of the unstructured texts were carried out. The number of synonyms, abbreviations, and misspellings that exist for terms decrease the performance of a classifier, thus, normalisation was performed. Pre-processing was carried out to reduce the document vector space. More precisely, the following measurements concerning text manipulation were carried out:

1. Normalisation of text and letter format
   - Case converter, to change capital letters into small letters; that improves keyword search since capital letters do not have to be considered anymore
   - Punctuation marks (e.g. , ; / </n>) are mostly erased or replaced with dots, so that the sentence structure is not ruined. Punctuation marks are not part of a text analysis, but it can be a hint for an unknown ADR or for class label 3. Therefore, the question mark ("?") is changed into the word *maybe* (Swedish: *kanske*)
2. Normalisation of words
   - Expansion to change abbreviations and acronyms to their full word (e.g. *pat* into *patient*)
   - Misspellings and synonyms
   For example, the following abbreviations and variations are found for the term *biverkning* (English: *side effect*): *biv, biverkn, biverkningarna*. For the term *biverkning* 44 misspellings were recognised.
3. Pre-processing
   - Number filter to reduce the vector space and increase computational speed
   - N-Char filter [N=2] to remove words that contain less than N letters
   - Stop Word Filter to filter out stop words (Leskovec et al., 2014)

3.4 Dictionaries and word lists

To support normalisation, three dictionaries were created:

*** Formatting Issue - Best Available Paper/Graphic ***

	Total	drug related ADR class 1	general ADR class 2	possible ADR class 3	no ADR class 4
Cefuroxim	1,243	11	840	134	258
Imovane	2,329	4	1,790	124	411
Oxynorm	2,886	16	2,120	143	607
Prednisolon	3,411	42	2,134	312	923
Primperan	4,882	34	3,237	296	1,315
Total	14,751	107	10,121	1,009	3,514
Share		0.7%	68.6%	6.8%	23.8%

Table 1: SEPR Drug Corpus with health records per class and drug

- misspellings (contains 460 words)
- abbreviations (contains 217 words)
- synonyms (contains 19 words)

These dictionaries were created by applying a Swedish grammar checker (Semushin, 2015) to a Bag of Word list. Marked words were either abbreviations or misspelling. The list was completed with corrected or expanded words. However, the dictionary for misspellings is not complete; the detection of misspellings was finalised after 460 words since the process was too time consuming.

To support the process of matching and filtering words or phrases, different word lists were created and can be found here [4]:

- **ADR terms** (12 words): they are found with an online thesaurus (SinovumMedia, 2015), online translator (bab.la, 2015), and lexica (Wikipedia, 2015; FASS, 2014b; Healthcare, 2015; Elsevier, 2015).
- **ADR phrases** (195 phrases): for this case a new iterative technique was developed: a term frequency counter for ADR terms (12 words) is applied. Words that co-occur with ADR terms are filtered and combined into one phrase. These three steps (applying a frequency counter, filter terms, and combining them) is executed three times in total. The results are in a list of ADR phrases, which is checked manually.
- **drugs** (5 words): the five chosen drugs
- **drug-related ADR symptoms**: ADR that are known, related to one of the five chosen drugs, and listed in FASS (2014b)
- **general ADR symptoms** (93 words): the list is generated with terms retrieved with the web-content-mining technique from FASS (2014b). The retrieval and extraction process was not part of this research study.

[4]http://dsv.su.se/health/dictionaries

The word list 'ADR terms' contains, for example: *reaktion, bieffekt, biverkning, iatrogen* (in English: *reaction, side effects, adverse events, iatrogenic*). The word list 'ADR phrases' contains, for example: *mycket vanlig biverkning, förmodligen biverkning, känd biverkning, misstanke om biverkning* (in English: *very common side effect, possible side effect, known side effect, suspected adverse reaction*)

Both dictionaries and word lists were reduced to their stem with the Swedish version of Snowball stemmer (Porter, 2001) to improve hit accuracy and to bypass spelling errors.

3.5 Matching and filtering

To reduce the word vector space even more, words and phrases were matched according to the created word lists and filtered from the health records. This match and filter process involves two dimensions: matching words that belong to three different intensities, and filtering the matched terms into three different levels.

An ADR is a drug-symptom relation and is usually indicated in a sentence with a combination of terms for drug, reaction (like *biverkning*), and symptom, such as *Cefuroxim gav biverkning* (English: *Cefuroxim gave ADR*), or *Patient har fått huvudvärk av Cefuroxim* (English: *Patient got headache from Cefuroxim*). A negated or suspected ADR may be indicated with a combination of a negated or suspected reaction, drug and (negated) symptom, for example *inga biverkningar* (English: *no ADR*), *ingen huvudvärk av Cefuroxim* (English: *no headache from Cefuroxim*). Thus, four types of terms were distinguished to match ADRs: reactions, drugs, symptoms, and help words. These terms were matched in the records in three different intensities to meet the variants of an ADR/no ADR/possible ADR that can be expressed and to investigate which matching intensity gains best prediction performance. The matching intensities (dimension) are:

1. Intensity 1 - **ADR terms**: terms that indicate an ADR (e.g. *biverkning, bieffekt, biverkan, läkemedelsutlöst*) were tagged as [Reaction]; additionally the names of the five chosen drugs were tagged as [Drug]. The tagged terms are mostly nouns.
2. Intensity 2 - **ADR terms plus help words**: helping words and terms that may indicate an ADR (e.g. *förmodligen, misstänkt*) or no-ADR (e.g. *ingen, utan*) are tagged as [HelpWord] additionally to intensity level 1. The additionally tagged terms are adjectives, adverbs, and verbs.
3. Intensity 3 - **ADR terms, help words, plus symptoms**: in addition to the two levels above, two types of symptoms were tagged: general symptoms that often are ADRs as [Symptoms], and additionally, symptoms that are ADR related to the drug as [Symptoms]. The additionally tagged terms are mostly nouns.

With matching and tagging a text is modified to: "Patient har inga[HelpWord] biverkningar [Reaction] av strålbehandlingen ännu, mår inte [HelpWord] illa [Symptom] och har inga [HelpWord] huvudvärk [Symptom]. Tar Cefuroxim [Drug]." (English: "The patient has no [HelpWord] adverse drug reactions [Reaction] of radiotherapy yet, do not [HelpWord] feel bad [Symptoms] and have no [HelpWord] headaches [Symptoms]. Taking Cefuroxime [Drug].")

Since words create their meaning in a context, a second dimension was chosen to filter the matched words in three different levels of filtering. For example, a term such as *adverse reaction* influences the class label depending on whether the term is part of *patient has no adverse reactions* or of *patient has an adverse reaction to Cefuroxim*. That is why the context of the matched term is considered with the filter level. Eriksson et al. (2013) applied designed dictionaries to filter ADR relevant compounds in clinical texts. However, the number of features should also be reduced without losing important information so that a machine learning algorithm runs in an appropriate time. Therefore, the following filter levels are defined:

1. Filter level A - **only matched words**: only the tagged words are filtered. This means that if the term *biverkning* (English: *ADR*) is tagged, it is filtered, no matter if the term is part of the sentence *ingen biverkning* (English: *no ADR*) or *pratar om biverkning* (English: *talking about ADR*); however, helping words like *ingen* (English: *no*) are also filtered. The disadvantage of only matching and filtering keywords is that the order of words is not considered nor are negations.
2. Filter level B - **matched words and their neighbours**: tagged words and their word neighbours in the sentence are filtered (N-grams on word level). The idea behind this is to find and include words in the classification model that are close by the keywords and may influence their meaning, but were not tagged on purpose. Still, neither the order of words nor negations are considered.
3. Filter level C - **phrases and tag as one attribute**: ADR phrases are filtered, in addition to tagged words depending on the intensity level. The document vector contains of both words and word groups, where a word represents one feature and an ADR phrase represents one feature. Here, negations and word order concerning ADR phrases are considered.

The reasons for this choice are, firstly, the attributes in a document vector usually consist of one word (as in level 1); however, the filtering should not be biased (therefore, level 2 was created). Secondly, a solution for considering negations and word order was needed (therefore, level 3 was defined). With the filtered words and phrases, a document vector is formed, and machine learning algorithms are applied.

3.6 Machine Learning

As mentioned in the Methods section, the dataset containing of five drugs and four classes (14,751 records) must be reduced to run a Machine Learning tool in an appropriate amount of time. Moreover, dividing the dataset into sub-datasets helps to distinguish different combinations of classes, and thus, a comparison for determining which combination yields higher prediction results. It also meets the aim of designing a flexible method that is able to distinguish between different kinds of ADRs. Therefore, three sub-datasets were created that contain:

1. Health records of patients that take either Imovane or Oxynorm and are class-labelled with 2 (general ADRs), 3 (maybe ADRs), or 4 (no ADRs) to distinguish patients that may have ADRs from patients that definitely have ADRs or no ADRs at all.
2. Health records of patients that take Prednisolon and are class-labelled as 1 (ADR related to Prednisolon) or 2 (general ADRs) to distinguish patients with Prednisolon related ADR from patients with ADRs related to other drugs.
3. Health records of patients that take Cefuroxim and are class-labelled with 1 (Cefuroxim related ADRs), 2 (general ADRs), 3 (maybe ADR), or 4 (no ADR).

Other combinations are conceivable and are planned for the future.

Decision tree (DT) (Quinlan, 1993), Random Forest (RF) (Breiman, 2001) and Library Support Vector Machine (LibSVM) (Chih-Chang and Lin, 2001) were chosen as supervised machine learning algorithms. The produced model is applied on a test dataset containing 30% of the instances gained with stratified sampling. The predicted classes of the test data instances are evaluated with precision *(P)*, recall *(R)*, and F-score as performance measurer.

Furthermore, as the machine learning tool KNIME (Berthold et al., 2007) was chosen with WEKA add-ons for decision tree, random forest and libSVM (KNIME, 2015).

4 Results

With this research project a method has been developed that is able to do both predict, if an ADR/possible ADR in a health record occur, and identify an ADR as a drug-symptom relation. The latter was realised with attribute selection. With the presented method there were two results: firstly, prediction performance measurements as presented in Table 2, and secondly, attributes that are selected and may reveal drug-symptom relationships.

4.1 Best results for classification

The best results for a classification on a test dataset that contains of 30% of the health records chosen with stratified sampling were achieved with libSVM (default parameters): precision of 0.69, recall of 0.66 and F-score of 0.67. The prediction was performed on a 3-class sub-dataset containing patients that take Imovane or Oxynorm, and health records that were class-labelled with 'general ADR' (class 2), 'possible ADR' (class 3), or 'no ADR' (class 4). 10-cross-fold validation was also carried but led to lower prediction performance.

4.2 Iterative filter and tag technique

A new iterative technique was introduced to filter ADR phrases, to combine them into one feature, and to build a document vector. This technique improves prediction performance for the best achieved result (3-class problem, intensity 1) from an F-score of 0.46 to an F-score of 0.65 (both with DT) and from an F-score = 0.44 to an F-score of 0.67 (with libSVM). This is an F-score improvement of 41% and 52%, respectively, for the 3-class problem, intensity 1.

4.3 Results for drug-symptom relation extraction/feature selection

With feature selection, the most important features concerning the class labels are selected. A selected attribute is important if it supports distinguishing one class from the other. Here, it means an attribute helps to distinguish patients that, for example, have an ADR from patients that have none. However, a selected term does not mean that it is an ADR.

For the sub-dataset that is reduced to patients taking the drug Prednisolon (2-class problem, intensity level 3, filter level C, F-score of 0.62), a classifier must distinguish patients that show Prednisolon-related ADR from patients with general ADR. With attribute selection, 36 attributes were considered important. Four of them were categorised as ADR phrases, three can be related to Prednisolon, five terms are mentioned in FASS (FASS, 2014b) as either known ADR of Prednisolon or symptoms that are treated with Prednisolon. Eight of the 37 selected attributes are tagged as symptoms, but they are not mentioned as ADRs of Prednisolon nor as a symptom treated with Prednisolon (FASS, 2014b). One of these symptoms is *body weight*, which is a known side effect of cortisone, however, it is not mentioned as side effect of Prednisolon. Sixteen attributes cannot be evaluated clearly.

For the sub-dataset that contains patients taking the drug Cefuroxim (4-class problem, intensity level 1, filter level C, F-score of 0.48), 13 attributes are considered important with attribute selection, and all belong to ADR phrases.

4.4 Term frequencies

The term frequencies of tagged ADR terms were compared in the health records. Synonyms for an ADR are not equally distributed across the classes. For example, in class 2 the word *biverkning* (English: *adverse reaction*) occurs 320 times per 100 health records, whereas in class 3, it occurs only 165 times per 100 health records. In class 3 the ADR term *biverkan* (English: *adverse effect*) is preferred with 139 occurrences compared to class 2 with 66 occurrences. For all three classes, the terms *biverkning* (English: *adverse reaction*), *reaktion* (English: *reaction*), *biverkan* (English: *adverse effect*), and *bieffekt* (English: *side effect*) are the preferred terms to describe an ADR. The term *överkänslig* (English: *hypersensitive*) is mentioned 63 times per 100 health records in classes 2 and 3, whereas in class 4 (no ADR) it is only mentioned 36 times per 100 health records.

*** Formatting Issue - Best Available Paper/Graphic ***

classification problem	filter level	intensity	libSVM P	libSVM R	libSVM F-score	DT P	DT R	DT F-score	RF P	RF R	RF F-score
4-class classification											
Cefuroxim	A	1	0.43	0.40	0.41	0.55	0.35	0.35	0.39	0.29	0.28
class labels 1,2,3,4	A	3							0.18	0.25	0.20
	C	1	0.17	1.00	0.20	0.50	0.55	0.51	0.45	0.33	0.34
	C	2	0.70	0.45	0.48				0.43	0.28	0.27
	C	3	0.40	0.46	0.35						
3-class classification											
Imovane & Oxynorm	A	1	0.50	0.43	0.44	0.61	0.45	0.46	0.25	0.33	0.28
class labels 2,3,4	A	3				0.59	0.48	0.51	0.25	0.33	0.28
	C	1	**0.69**	**0.66**	**0.67**	0.70	0.62	0.65	0.25	0.33	0.28
	C	3				0.66	0.57	0.60	0.25	0.33	0.28
2-class classification											
Prednisolon	A	1	0.60	0.57	0.58	0.49	0.50	0.49	0.49	0.50	0.49
class labels 1,2	A	2	0.74	0.54	0.56	0.99	0.54	0.57	0.49	0.50	0.49
	A	3	0.49	0.50	0.49	0.82	0.58	0.62	0.49	0.50	0.49
	C	1	0.69	0.57	0.61	0.99	0.54	0.57	0.49	0.50	0.49
	C	2	0.74	0.54	0.56	0.49	0.50	0.49	0.49	0.50	0.49
	C	3	0.49	0.50	0.49	0.99	0.54	0.57	0.49	0.50	0.49

Table 2: Performance measurer for a 30% test dataset

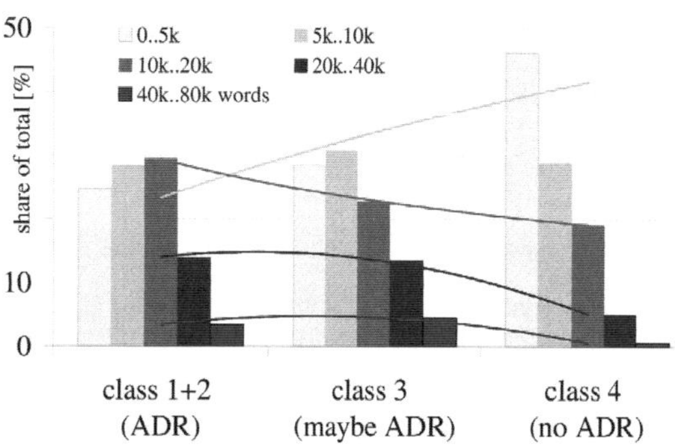

Figure 2: Distribution of record length per class.

4.5 Text length of health records

The length of a health record differs over the classes, as presented in Figure 2. The proportion of short texts shifts if patients are classified as 4 (no ADR), whereas health records classified as 1 or 2 (ADR occurred) tend to be longer.

5 Conclusions

Human beings are complex systems, so there can be great diversity in their reaction to a medical treatment. Clinical tests cannot carry out on all variants, even if all variants are known. Therefore, we have presented a method to perform ADR detection as post-marketing drug safety surveillance.

Supervised machine learning algorithms were applied on the SEPR Drug Corpus with 14,751 class labelled health records. A good prediction performance was yielded (F-score of 0.67, precision of 0.69 and a recall of 0.66). Hazlehurst et al. (2009) identified vaccine adverse effects with a supervised machine learning approach and found 181 of 319 true positives (57%). Aramaki et al. (2010) conducted machine learning and predicted 59% of the adverse drug events correctly. The present method finds 865 of 981 true positives (88.2%) in a 3-class dataset, which is an improvement of 49.5%.

The high ratio of misspelled terms for selected attributes calls for smart spell checks specialised for medical texts. The supervised machine learning algorithms prefer rare terms to distinguish classes. Misspellings disturb and mislead this process immensely.

The reason behind the different text length over the class labels (see figure 2) 1 to 4 may be that getting ADRs is just a question of time. Also, it may be that weakened, more sensitive, or ailing patients are prone to getting sick more often, and thus getting more medical treatment. Cascading effects arise, and therefore, it is more likely that an ADR occurs.

The fact that the word *överkänslig* (English: *hypersensitive*) occurs more often in health records of class 2 and 3 (139 occurrences vs 66 occurrences in class 4 per 100 health records) underscores our observation that patients with certain sensitivities are more likely to develop an ADR. If there is a correlation between ADRs and a patient's sensitivity, then this is even more of a rea-

*** Formatting Issue - Best Available Paper/Graphic ***

son to invest in post-marketing drug safety surveillance, since sensitive persons are limited in their participatiion in medical tests. The fact that the number of people that develop allergic reactions and other hypersensitivities has increased in recent years, highlights the urgency of post-marketing drug safety surveillance to better understand drug-symptom relations under special circumstances.

In the future, we plan to apply a spell check for Swedish, NER and parser techniques, to make the pre-processing faster and the prediction performance more accurate. Unfortunately, we used only one annotator; in the future, we will use at least two annotators to calculate the inter-annotation agreement (IAA). We also plan to test different methods of ADR expression extraction to perform machine learning and to obtain improved results.

Acknowledgements

The authors would like to thank Dr. Maria Kvist for her contributions as medical expert.

This work was partly supported by the project, High-Performance Data Mining for Drug Effect Detection, at Stockholm University, funded by Swedish Foundation for Strategic Research under grant IIS11-0053.

References

Helen Allvin, Elin Carlsson, Hercules Dalianis, Riitta Danielsson-Ojala, Vidas Daudaravicius, Martin Hassel, Dimitrios Kokkinakis, Heljä Lundgrén-Laine, Gunnar H Nilsson, Øystein Nytrø, Salanterä Sanna, Suominen Hanna, and Velupillai Sumithra. 2011. Characteristics of Finnish and Swedish intensive care nursing narratives: a comparative analysis to support the development of clinical language technologies. *Journal of Biomedical Semantics*, 2 Suppl 3:1–11.

Eiji Aramaki, Yasuhide Miura, Masatsugu Tonoike, Tomoko Ohkuma, Hiroshi Masuichi, Kayo Waki, and Kazuhiko Ohe. 2010. Extraction of adverse drug effects from clinical records. *Stud Health Technol Inform*, 160(Pt 1):739–743.

bab.la. 2015. Online dictionary for 27 languages. http://bab.la/. accessed 31/5/2015.

Michael R. Berthold, Nicolas Cebron, Fabian Dill, Thomas R. Gabriel, Tobias Kötter, Thorsten Meinl, Peter Ohl, Christoph Sieb, Kilian Thiel, and Bernd Wiswedel. 2007. KNIME: The konstanz information miner. In Christine Preisach, Hans Burkhardt, Lars Schmidt-Thieme, and Reinhold Decker, editors, *GfKl*, Studies in Classification, Data Analysis, and Knowledge Organization, pages 319–326. Springer.

Leo Breiman. 2001. Random forests. *Machine learning*, 45(1):5–32.

Chih-Chang and Chih-Jen Lin. 2001. *LIBSVM: a library for support vector machines*. http://www.csie.ntu.edu.tw/~cjlin/libsvm.

Hercules Dalianis, Martin Hassel, Aron Henriksson, and Maria Skeppstedt. 2012. Stockholm EPR Corpus: A Clinical Database Used to Improve Health Care. In *Swedish Language Technology Conference, SLTC-2012*, pages 17–18.

Hercules Dalianis, Aron Henriksson, Maria Kvist, Sumithra Velupillai, and Rebecka Weegar. 2015. HEALTH BANK - A workbench for data science applications in healthcare. In John Krogstie, Gustaf Juel-Skielse, and Vandana Kabilan, editors, *CAiSE Industry Track*, volume 1381 of *CEUR Workshop Proceedings*, pages 1–18. CEUR-WS.org.

Elsevier. 2015. Das Roche Lexikon Medizin. https://www.tk.de/rochelexikon/. accessed 31/5/2015.

Robert Eriksson, Peter Bjødstrup Jensen, Sune Frankild, Lars Juhl Jensen, and Søren Brunak. 2013. Dictionary construction and identification of possible adverse drug events in Danish clinical narrative text. *Journal of the American Medical Informatics Association*, 20(5):947–953.

FASS. 2014a. Fass atc register, http://www.fass.se/lif/atcregister?usertype=0. http://www.fass.se/LIF/atcregister?userType=0. accessed 31/5/2015.

FASS. 2014b. Läkemedelsindustriföreningens Service AB (LIF). Retrieved from 31/5/2015http://www.fass.se. accessed 31/5/2015.

U.S. Food Drug Administration FDA. 2014. *Inside Clinical Trials: Testing Medical Products in People*. http://www.fda.gov/Drugs/ResourcesForYou/Consumers/ucm143531.htm, accessed 31/5/2015.

Stefanie Friedrich. 2015. Detecting clinical entities using machine learning - How to find and predict (patients with) adverse drug reactions in medical records. Master thesis, Stockholm University.

Harsha Gurulingappa, Abdul Mateen Rajput, Angus Roberts, Juliane Fluck, Martin Hofmann-Apitius, and Luca Toldo. 2012. Development of a benchmark corpus to support the automatic extraction of drug-related adverse effects from medical case reports. *Journal of biomedical informatics*, 45(5):885–892.

*** Formatting Issue - Best Available Paper/Graphic ***

Brian Hazlehurst, Allison Naleway, and John Mullooly. 2009. Detecting possible vaccine adverse events in clinical notes of the electronic medical record. *Vaccine*, 27(14):2077–2083.

Alliance Healthcare. 2015. Gesundheit.de. http://www.gesundheit.de/. accessed 31/5/2015.

Rachel L Howard, Anthony J Avery, Serena Slavenburg, Simon Royal, G Pipe, Peter Lucassen, and Munir Pirmohamed. 2007. Which drugs cause preventable admissions to hospital? a systematic review. *British journal of clinical pharmacology*, 63(2):136–147.

KNIME. 2015. Knime weka data mining integration. https://www.knime.org/update/. accessed 31/5/2015.

Jurij Leskovec, Anand Rajaraman, and Jeffrey D. Ullman. 2014. *Mining of massive datasets*. Cambridge University Press, pub-CAMBRIDGE:adr, second edition.

Martin F Porter. 2001. Snowball: A language for stemming algorithms. http://snowball.tartarus.org.

David M. W. Powers. 2015. What the F-measure doesn't measure: Features, flaws, fallacies and fixes. *CoRR*, abs/1503.06410. http://arxiv.org/abs/1503.06410, accessed 31/5/2015.

J. Ross Quinlan. 1993. *C4.5: Programs for Machine Learning*. Morgan Kaufmann.

Cornelis Joost Van Rijsbergen. 1979. *Information Retrieval*. Butterworth-Heinemann.

Roland Roller and Mark Stevenson. 2014. Self-supervised relation extraction using umls. In *Information Access Evaluation. Multilinguality, Multimodality, and Interaction*, pages 116–127. Springer.

Sara Santiso, Alicia Pérez, Koldo Gojenola, IXA Taldea, Arantza Casillas, and Maite Oronoz. 2014. Adverse Drug Event prediction combining shallow analysis and machine learning. In *Proceedings of the 5th International Workshop on Health Text Mining and Information Analysis (Louhi)@ EACL*, pages 85–89.

Sergey Semushin. 2015. Notepad++ Spell-checking Plug-In, Swedish. https://github.com/Predelnik/DSpellCheck.git. accessed 31/5/2015.

SinovumMedia. 2015. Synonymer.se. http://www.synonymer.se/. accessed 31/5/2015.

Jürgen Stausberg and Joerg Hasford. 2011. Drug-related admissions and hospital-acquired adverse drug events in Germany: a longitudinal analysis from 2003 to 2007 of ICD-10-coded routine data. *BMC health services research*, 11(1):134.

Xiaoyan Wang, George Hripcsak, Marianthi Markatou, and Carol Friedman. 2009. Active computerized pharmacovigilance using natural language processing, statistics, and electronic health records: a feasibility study. *Journal of the American Medical Informatics Association*, 16(3):328–337.

Karin Wester, Anna K Jönsson, Olav Spigset, Henrik Druid, and Staffan Hägg. 2008. Incidence of fatal adverse drug reactions: a population based study. *British journal of clinical pharmacology*, 65(4):573–579.

WHO. 2015. Who collaborating centre for drug statistics methodology. http://www.whocc.no/atc/structure_and_principles/. accessed 31/5/2015.

Nandalal Wijesekera. 2013. *Documenting medical records - A handbook for doctors*. Health Information Systems Knowledge Hub School of Population Health,The University of Queensland, 1 edition.

Wikipedia. 2015. Adverse effect - Wikipedia, The Free Encyclopedia. https://en.wikipedia.org/wiki/Adverse_effect. accessed 31/5/2015.

NLP–Based Readability Assessment of Health–Related Texts: a Case Study on Italian Informed Consent Forms

Giulia Venturi◇**, Tommaso Bellandi**•**, Felice Dell'Orletta**◇**, Simonetta Montemagni**◇
◇Istituto di Linguistica Computazionale "Antonio Zampolli" (ILC–CNR)
ItaliaNLP Lab - *www.italianlp.it*
{name.surname}@ilc.cnr.it
•Laboratorio per le attivià di studio e ricerca applicata, Centro Gestione Rischio Clinico e Sicurezza dei Pazienti, Patient Safety Research Lab
bellandit@aou-careggi.toscana.it

Abstract

The paper illustrates the results of a case study aimed at investigating and enhancing the accessibility of Italian health–related documents by relying on advanced NLP techniques, with particular attention to informed consent forms. Results achieved show that the features automatically extracted from the linguistically annotated text and ranging across different levels of linguistic description have a high discriminative power in order to guarantee a reliable readability assessment.

1 Introduction

Within an information society, where everyone should be able to access all available information, improving access to written language is becoming more and more a central issue. This is the case of health–related information which should be accessible to all members of the society, including people who have reading difficulties e.g. as a result of a low education level, or of language–based learning disabilities, or because the language of the text is not their native language (WHO, 2015). It is a widely acknowledged fact that poor communication between physician and patients predisposes to medical malpractice cases (Kohn et al., 2000). Patient safety is a global challenge since the evidence on the burden of adverse events emerged in the past 15 years. An estimate of 43 millions adverse events occur in one year globally, with more than 50% of preventable events (Jha et al., 2013). In Italy, the incidence is of 5.2% on in–hospital admissions (Tartaglia et al., 2012) and the direct cost related to the prolongation of the stay are up to 3bln Euros in one year, roughly 3% of the funds of the National Healthcare Service (Albolino et al., 2013). The indirect costs related to claims is also high, amounting up to 1bln Euro in one year.

For all these reasons, the medical community has always shown strong interest in the improvement of health–related information in terms of document quality and understandability. Studies carried out so far mainly focused on traditional readability assessment methods, such as e.g. the Flesch–Kincaid measure (Kincaid, 1975) for the English language or the GulpEase index for Italian (Lucisano and Piemontese, 1988). According to them, the readability of medical texts is assessed by relying on basic text features such as sentence and word length, the only ones which could be automatically extracted from texts when these measures were originally conceived.

Recently, concerns have been raised about the effectiveness of traditional readability indices in capturing linguistic factors related to text complexity (Gemoets et al., 2004; Clerehan et al., 2005). This follows from the fact that now it is possible to carry out readability assessment against linguistically annotated texts, i.e. enriched with detailed and multi–level linguistic information generated by Natural Language Processing (NLP) components. Providing complex scientific information in a way that is comprehensible to a lay person is thus a challenge that nowadays can be addressed by deploying NLP techniques to capture a wide range of multi–level linguistic (e.g. lexical, syntactic, discourse) features and using statistical machine learning to build advanced readability assessment tools (Dell'Orletta et al., 2014a). So far, very few attempts have been devoted to the use of advanced NLP techniques to assess the readability of health–related texts; to our knowledge, none of them deals with Italian.

In this paper, we report the first results of a case study aimed at assessing the readability of a corpus of Italian informed consent forms on the basis of NLP–enabled surface, lexical and syntactic features. Among health–related texts, we focused on informed consent forms since ineffective doctor–

Proceedings of the Sixth International Workshop on Health Text Mining and Information Analysis (Louhi), pages 131–141,
Lisbon, Portugal, 17 September 2015.

patient communication is often due to a weak or lacking informed consent (Korenman et al., 2015).

The case study was carried out in the framework of the collaboration between the Institute of Computational Linguistics of the Italian National Research Council (ILC–CNR) and the Centre for Clinical Risk Management and Patient Safety (GRC) of the Tuscany region whose final goal is the development of advanced technologies to support the improvement of doctor–patient communication. In particular, it originates from the fact that in 2010 GRC was appointed to manage a communication and compensation program on adverse events, in order to improve the effectiveness and efficiency of claims management. Thanks to this programme, after 5 years the efficiency has strongly improved, with an estimate saving of 50millions Euro per year and a reduction of 5 months to close the claim. Yet, the number of claims is still stable and a recurrence of cases related to ineffective doctor–patient communication, often related to a weak or lacking informed consent, continues to be observed. The collaboration between ILC–CNR and GRC is aimed at creating the prerequisites for improvig the effectiveness of doctor–patient communication. This goal is pursued by designing and developing a writing tool for clinical practitioners which includes advanced functionalities for the evaluation of the quality of written documents and for supporting their simplification (whenever needed): the paper reports the results of preliminary investigations aimed at evaluating the readability of a wide corpus of documents presented to patients for informed consent, covering a wide range of clinical specialties and released by different healthcare trusts.

2 Background

It is a widely acknowledged fact that NLP techniques have an impact on the design of readability measures enabling to capture complex linguistic features with a significant gain in performance (François and Miltsakaki, 2012). However, differently from other application scenarios, little effort has been devoted so far in the biomedical domain to fully exploit NLP potentialities to evaluate the readability of health–related texts and to support clinical practitioners in the simplification, whenever needed, of the documents they produce. NLP–based readability assessment approaches reported so far for the biomedical domain differ with respect to: whether readability assessment is carried out as a classification task or in terms of ranking; the typology of features taken into account; the application within which readability assessment is carried out; and, last but not least, the language dealt with.

2.1 Methods and Features

Classification–based methods carry out readability assessment by assigning a given document to predefined readability classes: this is the case, for instance, of Kauchak et al. (2014) who built a machine learning classifier for predicting the difficulty of medical texts trained on a data set of aligned sentence pairs collected from English Wikipedia and Simple English Wikipedia. Interestingly, in the biomedical literature on readability assessment readability classes are typically restricted to two, i.e. easy vs. difficult. However, the main drawback of classification models is that they require training data, which may not exist, especially for a specific domain. An alternative to this method is represented by ranking–based approaches, positioning the document being analysed within a readability ranking scale: this approach is better suited for dealing with less resourced languages or to meet the needs of specific domains. In the biomedical domain, this method is adopted, among others, by: Kim et al. (2007), who developed a domain–specific approach to readability assessment calculating a distance score based on whether and to what extent text features of a test document differ from those of an easy sample (consisting in a collection of various web health information resources); or Zeng-Treitler et al. (2012) who, with the aim of improving the rank–based approach by Kim et al. (2007), used a wider set of lexical features also taking into account frequency information.

For what concerns the typology of features, NLP–based approaches proposed so far mainly focus on a combination of grammatical features, typically represented by the distribution of Parts–Of–Speech or of noun phrases, and lexical features, such as the distribution of domain terms with respect to domain–specific vocabularies, e.g. the Unified Medical Language System (UMLS) vocabulary. This is the case, e.g., of Proulx et al. (2013) who, by combining grammatical and vocabulary features, developed a tool specifically addressing the needs of clinicians and health educators for both readability assessment and enhancement. Since vocabulary plays a key role in health text readability, the most important extensions taken into account are concerned with lexical features. Starting from the assumption that more frequent terms are also easier to understand, Zeng-Treitler et al. (2012) included among the lexical features the distribution of terms with respect to two general–purpose resources, i.e. the Penn Treebank (Marcus et al., 1999) and the Google's Web 1T 5–gram Version 1 with n–gram frequency counts (Brants and Franz, 2006). For what concerns grammatical information, readability assessment in the biomedical domain does not go beyond to the distribution of Parts–Of–Speech and/or noun phrases: i.e. to our knowledge none of the domain–specific methods proposed so far makes use of syntactic features that can be extracted from the output of a syntactic parser.

2.2 Applications and Languages

Readability assessment is tackled from various perspectives with different applications in mind, giving rise to different tasks ranging from discerning easy vs. difficult electronic health records (Zeng-Treitler et al., 2007b), consumer health web sites, patient blogs and patient educational material (Leroy et al., 2006), to the simplification of medical texts carried out by devising metrics that can help making health–related documents more comprehensible to consumers. Due to the central role of lexical features in determining the readability of health–related texts, lexical simplification turned out to be the most explored level of text simplification. Different methods were devised to make health documents more comprehensible to consumers by reducing vocabulary difficulty. Even if with some differences, all approaches rely on the identification of difficult words and their replacement with easier synonym words. For this purpose, both domain–specific (e.g. Unified Medical Language System (UMLS), open–access collaborative (OAC), consumer health vocabulary (CHV)) and general–purpose (WordNet synonyms and hyperonyms, Wiktionary definitions, frequency counts of words in Google Web Corpus) resources were used. This is the case of Zeng-Treitler et al. (2007a) who built a prototype text translator to simplify narrative reports in electronic health reports, and of Leroy et al. (2012) who developed a semi–automatic algorithm tested on patient materials available on–line whose original and simplified version was presented for evaluation to a medical librarian (to measure the *perceived* difficulty) and to laymen (to measure the *actual* difficulty). Kandula et al. (2010) defined a text simplification method relying on both semantic and syntactic features: following Siddharthan (2006)'s approach, their algorithm is articulated into three steps, i.e. sentences longer than 10 words are first splitted, then Part–Of–Speech patterns are identified, and transformational rules are applied to generate shorter sentences.

Readability metrics developed so far typically deal with English, with few attempts tackling other languages. The most prominent exception is represented by Swedish, for which a quantitative corpus analysis of a collection of radiology reports was carried out as a preliminary step towards the development of a Swedish text simplification tool (Kvist and Velupillai, 2013). Similarly to English, simplification algorithms for Swedish health–related documents were devised by relying on synonym replacement methods (Abrahamsson et al., 2014), or on automatic detection of out–of–dictionary words and abbreviations, or on compound splitting and spelling correction (Grigonyte et al., 2014). Initiatives carried out so far for what concerns Italian are based on traditional readability formulas. This is the case of the ETHIC (*Evaluation Tool of Health Information for Consumers*) project (Cocchi et al., 2014), aimed at developing an effective tool for biomedical librarians and health information professionals to assess the quality of produced documents and to support them in preparing texts of increasing quality, suitable and comprehensible for patients and consumers in general. The tool carries out text readability and lexical understandability evaluation by resorting to the GulpEase readability formula (Lucisano and Piemontese, 1988) and the Basic Italian Vocabulary (De Mauro, 2000). Another relevant case study dealing with different languages also including Italian is reported in Terranova et al. (2012), whose aim was to assess and improve the quality and readability of informed consent forms used in cardiology. Although readability assessment was carried out with traditional readability formulas to guarantee comparability of results across languages, the main novelty of this study is that the simplification of Italian consent forms was guided by a preliminary version of READ–

IT (Dell'Orletta et al., 2011), the first NLP–based readability assessment tool for Italian.

3 The Approach

Our approach to the assessment of readability of Italian health–related texts combines NLP–enabled feature extraction and state–of–the–art machine learning algorithms. In this case study, we chose to exploit a general–purpose readability assessment tool, represented by READ–IT (Dell'Orletta et al., 2011)[1], the first NLP–based readability assessment tool for Italian which combines traditional raw text features with lexical, morpho–syntactic and syntactic information (see Section 3.2). In READ–IT, analysis of readability is modelled as a classification task. In particular, readability classification is binary, i.e. it is based on a training set consisting of two corpora representative of difficult– vs. easy–to–read texts. The easy–to–read training set is represented by *Due Parole* ("2Par"), a newspaper specifically written for an audience of adults with a rudimentary literacy level or with mild intellectual disabilities: the articles in *2Par* were written by Italian linguists expert in text simplification using a controlled language at the lexicon and sentence structure levels (Piemontese, 1996). For the selection of the difficult–to–read training set we opted for texts belonging to the same class, i.e. newspapers. In particular, we used the daily newspaper *La Repubblica* ("Rep"): even if widely read by many people in Italy, the national statistics on literacy skills report that 71% of the Italian people can hardly comprehend texts of medium difficulty such as the *Rep* articles.

Two other qualifying features of the READ–IT approach to readability assessment are worth reporting here, namely: *i)* readability is assessed by considering a wide range of linguistic characteristics automatically extracted from linguistically annotated texts, and *ii)* readability analysis is carried out at both document and sentence levels. As reported in section 2, readability assessment in the biomedical domain typically relies on linguistic features extracted from automatically PoS–tagged texts: instead, our approach also includes features extracted from syntactically (i.e. dependency) parsed texts, thus making it possible to monitor a wider variety of factors affecting the readability of a text. The set of features can be parameterized creating the prerequisites for specializing the readability assessment measure with respect to different target audiences, specific domains of knowledge or with respect the type of textual object, i.e. the document or individual sentences. Assessing readability at both document and sentence levels allows highlighting specific text portions which require reformulation with respect to the used vocabulary or to the grammatical structure (Dell'Orletta et al., 2014c). In fact, similarly to other application scenarios, also in the biomedical domain evaluating the readability of individual sentences represents an essential prerequisite for text simplification, to be carried out at both lexical and syntactic levels (Kandula et al., 2010). Despite that sentence readability assessment is a qualifying feature of READ–IT, in what follows we will focus on document readability only.

For the experiments reported in the paper, we used general purpose readability models trained on newspaper corpora. This was an unavoidable choice, due to the lack of domain–specific resources annotated with grade levels to be used as training data. Although this makes achieved results still preliminary, it was a way to test effectiveness and reliability of the method on health–related texts. For what concerns the evaluation of achieved readability assessment results, the target readers of *2Par* (i.e. the READ–IT easy–to–read pole) were taken as coinciding with the target reader of health–related texts: the underlying assumption is that the informed consents classified as difficult–to read for the *2Par* low literacy readers are really complex and need to be simplified. Obviously, when a version of READ–IT specialized for the biomedical domain will be released, a qualitative evaluation of results will be needed. Previous studies resorted to the Cloze test to validate the reliability of their results or integrated editing capabilities into the developed tools in order to receive feedback from end users. The work carried out by Kandula and Zeng-Treitler (2008) represents an exception. They assembled a panel of experts to evaluate the readability of 324 different typology of English health documents: the rated collection was meant to be used as a gold standard to evaluate readability metrics.

3.1 The corpus

For this case study, we collected a corpus of 583 documents, for a total of 607,677 word tokens,

[1] http://www.italianlp.it/demo/read-it/

Features	*2Par*	*2IC*	*Rep*
Average sentence length	19.20	16.06	26.54
Average word length	4.98	6.75	5.18
% of lemmas (types) in BIV	74.58	57.24	67.09
% of lemmas (types) NOT in BIV	25.42	42.76	32.91
Type/token ratio (first 100 tokens)	0.55	0.72	0.72
Distribution of Parts–Of–Speech:			
– nouns	29.30%	28.51%	27.19%
– verbs	13.66%	11.83%	12.89%
– adjectives	5.92%	9.26%	6.40%
– prepositions	15.28%	16.19%	16.41%
Noun/verb ratio	2.14	2.41	2.11
Average length of the longest dependency link	7.91	6.43	10.28
Average parse tree depth	5.29	4.86	6.51
Average depth of embedded complement 'chains'	1.24	1.31	1.34
Distribution of 'chains' by depth:			
– 1 embedded complement	79.40%	74.25%	72.32%
– 2 embedded complements	17.02%	21%	21.42%
$\geq$ 3 embedded complements	3.27%	4.73%	5.87%
Main vs subordinate clauses distribution:			
– main clauses			
– subordinate clauses	26.14%	25.30%	32.36%
Average clause length	9.81	11.29	10.12
Distribution of verbal roots with explicit subject	74.69%	57%	64.30%

Table 1: Selection of linguistic features strongly characterizing the *2IC* corpus.

constituted by the procedures and the documents for informed consents currently used in all the 16 healthcare trust of the Regional Healthcare Service (RHS) of Tuscany, namely 4 academic hospitals and 12 local healthcare authorities. The documents were partitioned into different groups, classified according to the clinical specialty and the document type (procedure or user guide). Henceforth, we will refer to this corpus as the "Italian Informed Consent Corpus" (2IC).

Table 1 reports a selection of linguistic features which turned out to strongly characterize the *2IC* corpus with respect to the journalistic *2Par* and *Rep* corpora. This analysis is meant to compare domain–specific (i.e. biomedical) and general purpose corpora with the final aim of detecting the main linguistic features characterizing the language used in informed consent forms. The features were extracted from the corpus automatically tagged by the part–of–speech tagger described in Dell'Orletta (2009) and dependency–parsed by the DeSR parser (Attardi, 2006).

Starting from raw textual features, it can be noticed that the *2IC* corpus is characterized by shorter sentences (calculated as the average number of words per sentence) and longer words (calculated as the average number of characters per word) if compared with the *2Par* and *Rep* corpora. Starting from the assumption underlying traditional readability formulas assuming that longer sentences are more grammatically complex than shorter ones and that longer words are less comprehensible than shorter ones, this result witnesses the efforts of the authors of informed consents towards the use of an unavoidably complex vocabulary used, however, in simpler syntactic constructions. Interestingly enough, this is confirmed by the values of lexical features. Among them, it is worth noting that with respect to both *2Par* and *Rep* informed consents contain quite a lower percentage of lemmas (types) belonging to the "Basic Italian Vocabulary" (De Mauro, 2000), marked as BIV in Table 1 and corresponding to a list of 7000 words highly familiar to native speakers of Italian. This is in line with the outcomes of the studies on the discriminative power of vocabulary clues in readability assessment (see, among others, Petersen and Ostendorf (2009)). Obviously, this also reveals the massive use of health–related words specific to this domain of knowledge and here still considered as out–of–vocabulary lemmas. In addition, *2IC* texts show a higher Type–Token Ratio (TTR) value (which is computed for the first 100 tokens of each document), meaning that this text type is much richer lexically, with values which are closer to what observed with respect to *Rep*, here considered as representative of the class of difficult–to–read texts.

Consider now the distribution of Parts–Of–Speech across the *2Par*, *Rep* and *2IC* corpora. Informed consents are characterized by a high percentage of adjectives, prepositions and nouns, and by a low percentage of verbs: this gives rise to a much higher noun/verb ratio. According to Biber (1993), such different distributions represent significant dimensions of variation across textual genres. In particular, the higher noun/verb ratio reveals that informed consent forms are more informative than newspaper articles (Biber and Conrad, 2009), while the higher occurrence of nouns and prepositions is strongly connected with their presence within embedded complement 'chains' governed by a nominal head and including either prepositional complements or nominal and adjectival modifiers. Similarly to *Rep* articles, health–related documents contain a high percentage of complex nominal constructions (*Average depth of embedded complement 'chains'* in Table 1) with deep sequences of embedded complements. This is also reflected at the level of the probability distribution of embedded complement 'chains' by depth: if on the one hand we observe a lower percentage of short sequences (i.e. with depth=1) with respect to *2Par* here taken as representative of easy–to–read texts, on the other hand – similarly to *Rep* – a higher percentage of longer sequences (i.e. with depth=2 and =$\geq$ 3) is recorded.

Interestingly, however, besides complexity features such as "heavy" nominal constructions – possibly due to multi–word terminology – in informed consents low values are recorded for syntactic features typically associated with *structural* complexity, such as: parse tree depth, calculated in terms of the longest path from the root of the dependency tree to some leaf; length of dependency links, measured in terms of the words occurring between the syntactic head and the dependent; or the distribution of main vs. subordinate clauses. From this, we can conclude that language complexity in informed consent forms mainly lies at the level of *local* features of the parse tree. Other peculiar syntactic features of informed consents with respect to *2Par* and *Rep* are represented by longer clauses (*Average clause length* in Table 1, calculated as the average number of tokens per clause), and a lower percentage of verbal roots with explicit subject (calculated with respect to the total amount of verbal roots). For what concerns the latter, even if Italian is a pro–drop language, sentences characterized by elliptical constructions (e.g. verbal roots with explicit subjects) make a text more difficult–to–read and need to be simplified, as suggested in Barlacchi and Tonelli (2013).

3.2 Readability Assessment

For readability classification experiments we used READ–IT, the first NLP–based readability assessment tool devised for Italian. It operates on syntactically (i.e. dependency) parsed texts and assigns to each considered reading object - either a document or a sentence - a score quantifying its readability. READ–IT is a classifier based on Support Vector Machines using LIBSVM (Chang and Lin, 2001) that, given a set of features and a training corpus, creates a statistical model using the feature statistics extracted from the training corpus. Such a model is used in the assessment of readability of unseen documents or sentences. The assigned readability level ranges between 0 (easy–to–read) and 100 (difficult–to–read) referring to the percentage probability for the unseen documents or sentences to belong to the class of difficult–to–read documents. The score assigned by READ–IT can thus be seen as a score of text difficulty.

As fully described by Dell'Orletta et al. (2011), the tool is trained on *2Par* (taken as representative of the class easy–to–read texts) and on *Rep* (representing the class of difficult–to–read texts) articles and it exploits the wide typology of raw text, lexical, morpho–syntactic and syntactic features summarized in Table 2. This proposed four–fold partition of features closely follows the different levels of linguistic analysis automatically carried out on the text being evaluated, i.e. tokenization, lemmatization, morpho–syntactic tagging and dependency parsing. Such a partition was meant to identify those easy to extract features with high discriminative power in order to reduce the linguistic pre–processing of texts guaranteeing at the same time a reliable readability assessment.

The set of features used to build the statistical model can be parameterized through a configuration file. This creates the prerequisites for customising the readability assessment measure with respect to the target audience or to the sublanguage of a specific domain. According to the different types of features considered, READ–IT assigns different readability scores using the following four feature models:

1. **Base Model**, relying on *raw text* features only;

Feature category	Name
Raw Text	Average number of words per sentence Average number of characters per word
Lexical	Type/Token Ratio Lexical density *Basic Italian Vocabulary (BIV)* (De Mauro, 2000) rate
Morpho–syntactic	Part-Of-Speech unigrams Mood, tense and person of verbs
Syntactic	Distribution of dependency types Depth of the whole parse tree Average depth of embedded complement 'chains' Distribution of embedded complement 'chains' by depth Number of verbal roots Arity of verbal predicates Distribution of verbal predicates by arity Distribution of subordinate vs main clauses Relative ordering with respect to the main clause the Average depth of 'chains' of embedded subordinate clauses Distribution of embedded subordinate clauses 'chains' by depth Length of dependency links feature

Table 2: Linguistic features used for readability assessment purposes.

2. **Lexical Model**, relying on a combination of *raw text* and *lexical* features;

3. **Syntax Model**, relying on *morpho–syntactic* and *syntactic* features;

4. **Global Model**, combining all feature types, namely *raw text*, *lexical*, *morpho–syntactic* and *syntactic* features.

4 Results and Discussion

In this section, we discuss the outcome of the readability assessment experiments carried out on the *2IC* corpus described in Section 3.1. In order to identify the contribution of the different types of features in the assessment of the readability of informed consents, we focus on the results obtained by the base, lexical and syntactic READ–IT models (see, respectively, columns *Base*, *Lexical* and *Syntax* in Table 3). In what follows, we will focus on the results of the readability experiments carried out at the document level while an in depth investigation of the linguistic aspects affecting sentence readability is part of an on–going study.

Table 3 reports the results obtained with respect to the whole corpus, for all the 29 medical specialties; a score for each of the 4 considered macro–specialties (namely, *Surgery, Internal Medicine, Prevention and Medical Services*) is also computed, as the average of the scores recorded for each specialty. It can be noted that the whole corpus is characterized by a low readability level, even if with significant differences among the different readability models and across macro–specialties. Interestingly, the results obtained by the *Base* model show how raw text features such as sentence and word length are not really effective to capture the difficulty of these texts as well as the differences among them. This model can be seen as an approximation of the GulpEase index (Lucisano and Piemontese, 1988), i.e. the most used traditional readability measure for Italian which is based on the same raw text features (i.e. sentence and word length). This naturally follows from the results illustrated in Section 3.1, investigating the linguistic features characterizing *2IC* with respect to general purpose corpora: as Table 1 shows, *2IC* contains quite short sentences, a raw text feature typical of easy–to–read texts.

By comparing the scores obtained for the macro–specialties, it is worth noting that the score obtained with the *Base* model for the *Prevention* area is misleading: i.e, the prevention forms result to be more difficult than the *Internal Medicine* documents and only slightly easier than the *Medical Services* ones. The situation looks quite different if we consider instead the *Lexical* and the *Syntax* models: we can observe that the *Prevention* documents are easier–to–read than the documents of the other macro–specialties. On the contrary, sharp differences among the 4 macro–specialties and the 29 specialties occur as far as the *Lexical* and the *Syntax* models are concerned. In particular, all specialties turned out to be more difficult at the lexical than at the syntactic level. For what concerns the former, this follows from the high percentage of out–of–vocabulary lemmas characterizing the informed consents with respect to the Basic Italian Vocabulary: as expected, *Prevention* documents represent an exception, being the easiest–to–read macro–specialty at the lexical level.

Consider now the results obtained with the *Syntax* model, according to which all informed consents turned out to be less difficult–to–read with respect to the lexical level. As discussed in Section 3.1, the typology of features contributing to this result is related to *local* aspects of the parse tree, taken in literature as an index of language complexity, rather than to *structural* complexity features. This type of evidence will be used in the near future to customize the set of features to be taken into account in the construction of a domain–specific version of the *Syntax* readability model. Also in this case, *Prevention* documents turned out to be more readable than the other specialties.

5 Conclusion and Future Work

In this paper, we illustrated the preliminary but encouraging results of a broader and long–term study devoted to enhance the accessibility of Italian health–related documents by relying on advanced Natural Language Processing techniques: the case study reported in the paper focuses on informed consent forms, which play a key role in doctor–patient communication. For this purpose, we used READ–IT, a general purpose NLP–based readability assessment tool for Italian. The results obtained so far show that the features automatically extracted from the linguistically annotated text and ranging across different levels of linguistic description, also including syntax, have a high discriminative power to guarantee a reliable readability assessment. To our knowledge, this is the first application of an advanced NLP–based methodology for readability assessment of Italian health–related documents. The proposed methodology was tested on a corpus of Italian informed consents currently used in healthcare trusts of the Regional Healthcare Service of Tuscany.

The results obtained by comparing readability scores across the considered medical specialties with respect to the different READ–IT models revealed that – generally speaking – informed consents are more difficult–to–read at the lexical level than at the syntactic level. This is in line with the linguistic profiling results discussed in Section 3.1, according to which the *2IC* corpus contains a higher percentage of out–of–vocabulary words, even higher than difficult–to–read texts (i.e. *Rep*). Behind this general trend, significant differences are reported for the different specialties, e.g. the *Prevention* documents turned out to be easier–to–read than the documents of the other (macro–)specialties.

The higher difficulty recorded at the lexical level suggests that the general purpose READ–IT tool needs to be specialized at the level of the permitted vocabulary, which should also include a selection of basic domain terms to be used in informed consent forms without any penalization at the level of the readability score. We are already working in this direction. Two experts in healthcare quality assessment are currently evaluating the out–of–vocabulary lemmas automatically extracted from the *2IC* corpus by the $T2K^2$ (Text–to–Knowledge) platform (Dell'Orletta et al., 2014c) with the final aim of creating a domain–specific lexicon to be used in the specialized version of READ–IT we are currently developing. The lexicon will be internally organized into three classes of *i)* "domain–specific words", i.e. words that cannot be avoided within health–related documents (e.g. *anestesia* 'anesthesia'), *ii)* "technical words", i.e. words that are specific to the domain but that should be explained with a gloss in order to be fully understood by laymen (e.g. *complicanza* 'complication'), and *iii)* "technicalities", i.e. words that are used by experts but that should be replaced with a simpler synonym in order to be fully understood by laymen (e.g. *fistola* 'fistula'). Obviously, as suggested above the specialization will also be concerned with grammatical features.

From a more general perspective, these preliminary results show a severe lack of knowledge and skills on the design of readable informed consents within healthcare services. Clearly, we can interpret these findings in the bureaucratic framework within which the documents are produced, missing the goal of informing patients while accomplishing the legal duty to have a "piece of paper" reporting the signatures of doctors and patient in the healthcare record, without a clear explanation of the treatments. Further research is needed to design and evaluate systems to support the preparation of the documents of informed consent: in this context, the customization of the READ–IT

Medical Specialty	nº documents	nº tokens	READ–IT		
			Base	**Lexical**	**Syntax**
Anesthesiology	20	21,065	50	93.37	69.62
Colorectal surgery	2	1,997	75.18	100	93.81
Obesity surgery	3	8,091	51.63	93.42	59.20
General surgery	19	11,588	43.03	78.29	58
Plastic surgery	4	3,550	88.95	98.72	96.51
Thoracic surgery	9	5,608	94.98	99.94	95.55
Vascular surgery	16	22,739	88.64	98.13	97.62
Ophthalmology	7	10,496	49.21	98.89	61.29
Otorhinolaryngology	134	194,421	25.14	94.90	69.42
Orthopaedics	44	76,712	50.54	97.58	89.66
Obstetrics and gynecology	35	31,243	60.37	97.31	58.52
Urology	17	19,576	85.40	98.08	89.16
TOTAL: Surgery	**313**	**407,086**	**63.59**	**95.72**	**78.19**
Cardiology	54	39,887	66.20	94.50	78.99
Diabetology	1	297	23.05	100	45.68
Gastroenterology	9	9,856	41.12	87.90	59.82
Neurology	8	5,199	69.44	97.96	94.98
Oncology	3	1,692	46.34	99.73	96.07
Pulmonology	4	3,220	49.57	98.18	78.27
Senology	17	20,455	85.09	99.68	93.88
TOTAL: Internal Medicine	**96**	**80,309**	**54.26**	**96.85**	**78.24**
Psychology	13	11,651	80.44	96.25	98.32
Screening	8	2,007	53.13	65.14	50.60
Vaccine	1	2,852	33.72	100	71.76
TOTAL: Prevention	**22**	**16,510**	**55.76**	**87.13**	**73.56**
Genetics	11	6,416	56.26	95.65	81.45
Immunohematology and transfusion	43	45,962	56.84	93.39	83.47
Nuclear medicine	29	18,045	52.62	96.56	68.48
Radiology	24	17,358	63.78	98.61	78.68
TOTAL: Medical Services	**107**	**87,781**	**57.38**	**96.05**	**78.02**
General	33	8,928	51.59	87.81	88.27
Pediatrics	13	6,092	49.84	99.46	74.67
Rehabilitation	2	674	63.84	99.99	96.25

Table 3: Readability assessment results by the *Base*, *Lexical* and *Syntax* models organized by medical specialties.

tool will play a key role. A specialized version of READ–IT will be possibly integrated within the Electronic Patient Record, so that the informed consent becomes part of a process of shared decision making where the doctors prepare a readable message for the patient at the time of the decision for a clinical procedure and collect questions and comments, that in turn feebacks into a software capable to learn from the daily practice. A limitation of this approach is the exclusive reliance on written documents, while according to the current debate (Korenman, 2015) in ethics and medico-legal issues the informed consent should be the result of a process of communication where the written document supports the doctor–patient communication. Bringing this to an extreme perspective, the informed consent could be simply the transcription of the dialogue that demonstrates the provision of comprehensive information on the possible treatments for a disease and the shared decision on the best alternative for the involved parts. However, even in this futuristic scenario NLP technologies could play a role.

References

S. Albolino, T. Bellandi, R. Tartaglia, and A. Biggeri. 2013. The incidence of adverse events in tuscany: results from a regional study involving 36 hospitals. *Proceedings of ISQUA 30th International Conference*, 13–16 October, Edinburgh.

E. Abrahamsson, T. Forni, M. Skeppstedt, and M.Kvist. 2014. Medical text simplification using synonym replacement: Adapting assessment of word difficulty to a compounding language. *Proceedings of the 3rd Workshop on Predicting and Improving Text Readability for Target Reader Populations (PITR 2014, EACL Workshop)*, 57–65.

G. Attardi. 2006. Experiments with a multilanguage non-projective dependency parser. *Proceed-*

ings of the Tenth Conference on Computational Natural Language Learning (CoNLL-X '06), 166–170.

G. Barlacchi and S. Tonelli. 2013. ERNESTA: A Sentence Simplification Tool for Children's Stories in Italian. *Proceedings of the 14th Conference on Intelligent Text Processing and Computational Linguistics (CICLing-2013)*, 476–487.

D. Biber. 1993. Using Register–diversified Corpora for General Language Studies. *Computational Linguistics Journal*, 19(2): 219–241.

D. Biber and S. Conrad. 2009. Genre, Register, Style. Cambridge: CUP.

T. Brants and A. Franz. 2006. *Web 1T 5–gram Version 1*. Linguistic Data Consortium, Philadelphia.

C. C. Chang and C. J. Lin. 2001. *LIBSVM: a library for support vector machines*. Software available at `http://www.csie.ntu.edu.tw/~cjlin/libsvm`

R. Clerehan R. Buchbinder, and J. Moodie. 2005. A linguistic framework for assessing the quality of written patient information: Its use in assessing methotrexate information for rheumatoid arthritis. *Health Education Research*, 20(3):334–344.

S. Cocchi, M. Mazzocut, C. Cipolat Mis, I. Truccolo, E. Cervi, R. Iori, and D. Orlandini. 2014. ETHIC. Evaluation Tool of Health Information for Consumers. Development, features and validation. *Divided we fall, united we inform. Building alliances for a new European cooperation, 14th EAHIL Annual Conference*, Roma (Italy), 11-13 June.

T. De Mauro. 2000. *Il dizionario della lingua italiana*. Torino, Paravia.

F. Dell'Orletta. 2009. Ensemble system for Part-of-Speech tagging. *Proceedings of Evalita'09, Evaluation of NLP and Speech Tools for Italian*, Reggio Emilia, December.

F. Dell'Orletta, S. Montemagni, and G. Venturi. 2011 READ–IT: assessing readability of Italian texts with a view to text simplification. *Proceedings of the Second Workshop on Speech and Language Processing for Assistive Technologies (SLPAT)*, Edinburgh, UK, 73–83.

F. Dell'Orletta, S. Montemagni, and G. Venturi. 2014a. Assessing document and sentence readability in less resourced languages and across textual genres. *Recent Advances in Automatic Readability Assessment and Text Simplification. Special issue of International Journal of Applied Linguistics*, 165:2, John Benjamins Publishing Company, 163–193.

F. Dell'Orletta, M. Wieling, A. Cimino, G. Venturi, and S. Montemagni. 2014b Assessing the Readability of Sentences: Which Corpora and Features? *Proceedings of 9th Workshop on Innovative Use of NLP for Building Educational Applications (BEA 2014)*, Baltimore, Maryland, USA, 163–173.

F. Dell'Orletta, G. Venturi, A. Cimino, S. Montemagni. 2014c. T2K: a System for Automatically Extracting and Organizing Knowledge from Texts. In *Proceedings of 9th Edition of International Conference on Language Resources and Evaluation (LREC 2014)*, pp. 2062–2070, 26-31 May, Reykjavik, Iceland.

T. François and Eleni Miltsakaki. 2012. Do NLP and machine learning improve traditional readability formulas? *Proceedings of the NAACL-HLT 2012 Workshop on Predicting and Improving Text Readability for target reader populations (PITR 2012)*, 49–57.

A.K. Jha, I. Larizgoitia, C. Audera-Lopez, N. Prasopa-Plaizier, H. Waters, and D. Bates. 2013. The global burden of unsafe medical care: analytic modelling of observational studies. *BMJ Quality and Safety*, 22(10):809–15.

S. Kandula, D. Curtis, and Q. Zeng-Treitler. 2010. A semantic and syntactic text simplification tool for health content. *Proceedings of the American Medical Informatics Association Annual Symposium*, United States:366–70.

S. Kandula and Q. Zeng-Treitler. 2008. Creating a Gold Standard for the Readability Measurement of Health Texts. *Proceedings of the American Medical Informatics Association Annual Symposium*, Washington, DC, USA, 353–357.

D. Kauchak, O. Mouradi, C. Pentoney, and G. Leroy. 2014. Text simplification tools: Using machine learning to discover features that identify difficult text. *Proceedings of the 47th Hawaii International Conference on System Sciences (HICSS)*, Waikaloa, Big Island, Hawaii:2616–2625.

H. Kim, S. Goryachev, G. Rosemblat, A. Browne, A. Keselman, and Q. Zeng-Treitler. 2007. Beyond Surface Characteristics: A New Health Text-Specific Readability Measurement. *Proceedings of the American Medical Informatics Association Annual Symposium*, 418–422.

J. P. Kincaid, L. R. P. Fishburne, R. L. Rogers and B. S. Chissom. 1975. *Derivation of new readability formulas for Navy enlisted personnel.* Research Branch Report, Millington, TN: Chief of Naval Training, pp. 8–75.

Korenman S. 2015. *Enduring and emerging challenges of informed consent.* N Engl J Med. 2015 May 28;372(22):2171-2.

M. Kvist and S. Velupillai. 2013. Professional Language in Swedish Radiology Reports – Characterization for Patient-Adapted Text Simplification. *Proceedings of the Scandinavian Conference on Health Informatics 2013*, Copenhagen, Denmark:55–59.

L. T. Kohn, J. M. Corrigan, D. S. Donaldson. 2000. *To err is human: building a safer health system.* Washington, DC: National Academy Press.

2015. Enduring and emerging challenges of informed consent. *New England Journal of Medicine*, 372(22):2171-2.

D. Gemoets, G. Rosemblat, T. Tse, and R. Logan. 2004. Assessing Readability of Consumer Health Information: An Exploratory Study. *Medinfo*, 868–873.

G. Grigonytė, M. Kvist, S. Velupillai, and M. Wirén. 2014. Improving Readability of Swedish Electronic Health Records through Lexical Simplification: First Results. *Proceedings of the 3rd Workshop on Predicting and Improving Text Readability for Target Reader Populations (PITR 2014, EACL Workshop)*, 74–83.

G. Leroy, E. Eryilmaz, and B.T. Laroya. 2007. Health information text characteristics. *Proceedings of the American Medical Informatics Association Annual Symposium*, Washington DC:479–483.

G. Leroy and J. E. Endicott. 2012. Combining NLP with Evidence-based Methods to Find Text Metrics Related to Perceived and Actual Text Difficulty. *Proceedings of the 2Nd ACM SIGHIT International Health Informatics Symposium*, Miami, Florida, USA:749–754.

G. Leroy, J. E Endicott, O. Mouradi, D. Kauchak, and M. L. Just. 2012. Improving Perceived and Actual Text Difficulty for Health Information Consumers using Semi-Automated Methods. *Proceedings of the American Medical Informatics Association Annual Symposium*, 522–531.

P. Lucisano and M. E. Piemontese. 1988. *Gulpease. Una formula per la predizione della difficoltà dei testi in lingua italiana.* In Scuola e Città (3), pp. 57–68.

M. P. Marcus, M. A. Marcinkiewicz, and B. Santorini. 1999. Building a large annotated corpus of English: The Penn Treebank. *Computational linguistics*, 19(2), MIT Press:313-330.

M. E. Piemontese. 1996. *Capire e farsi capire. Teorie e tecniche della scrittura controllata* Napoli, Tecnodid.

S. E. Petersen and M. Ostendorf. 2009. A machine learning approach to reading level assessment. In *Computer Speech and Language (23)*, 89–106.

J. Proulx, S. Kandula, B. Hill, and Q. Zeng-Treitler. 2013. Creating Consumer Friendly Health Content: Implementing and Testing a Readability Diagnosis and Enhancement Tool. *Proceedings of the 46th Hawaii International International Conference on-Systems Science (HICSS- 46 2013)*, 2445–2453.

A. Siddharthan. 2006. Syntactic Simplification and Text Cohesion. *Research on Language and Computation*, Volume 4, Issue 1, Springer Science, the Netherlands:77–109.

R. Tartaglia, S. Albolino, T. Bellandi, E. Bianchini, A. Biggeri, G. Fabbro, L. Bevilacqua, A. Dell'erba, G. Privitera, and L. Sommella. 2012. Eventi avversi e conseguenze prevenibili: studio retrospettivo in cinque grandi ospedali italiani. *Epidemiologia & Prevenzione*, 36(3-4):151–61.

G. Terranova, M. Ferro, C. Carpeggiani, V. Recchia, L. Braga, R. Semelka, and E. Picano. 2012. Low Quality and Lack of Clarity of Current Informed Consent Forms in Cardiology - How to Improve Them. *Journal of the American College of Cardiology (JACC): Cardiovascular Imaging*, Elsevier inc., vol. 5(6):649–655.

World Health Organization. 2015. WHO global strategy on people–centred and integrated health services. Interim Report available at `http://apps.who.int/iris/bitstream/10665/155002/1/WHO_HIS_SDS_2015.6_eng.pdf`

Q. Zeng-Treitler, H. Kim, S. Goryachev, A. Keselman, L. Slaughter, and C. Smith. 2007b. Text characteristics of clinical reports and their implications for the readability of personal health records. *Studies in Health Technology and Informatics*, 129(2):1117–1121.

Q. Zeng-Treitler, S. Goryachev, H. Kim, A. Keselman, and D. Rosendale. 2007a. Making Texts in Electronic Health Records Comprehensible to Consumers: A Prototype Translator. *Proceedings of the American Medical Informatics Association Annual Symposium*, 846–850.

Q. Zeng-Treitler, S. Kandula, H. Kim, and B. Hill. 2012. A Method to Estimate Readibility of Health Content. *Proceedings of the ACM SIGKDD Workshop on Health Informatics (HI-KDD 2012)*, Beijing, China.

Mining and Ranking Biomedical Synonym Candidates from Wikipedia

Abhyuday N Jagannatha
College of Information and Computer Sciences,
University of Massachusetts, Amherst
MA 01003, USA
abhyuday@cs.umass.edu

Jinying Chen
Department of Quantitative Health Sciences,
University of Massachusetts, Worcester
MA 01605, USA
jinying.chen@umass-med.edu

Hong Yu
Veterans Administrative Medical Center,
Bedford
MA 01730, USA
hong.yu@umassmed.edu

Abstract

Biomedical synonyms are important resources for Natural Language Processing in Biomedical domain. Existing synonym resources (e.g., the UMLS) are not complete. Manual efforts for expanding and enriching these resources are prohibitively expensive. We therefore develop and evaluate approaches for automated synonym extraction from Wikipedia. Using the inter-wiki links, we extracted the candidate synonyms (anchor-text e.g., "increased thirst") in a Wikipedia page and the title (e.g., "polyuria") of its corresponding linked page. We rank synonym candidates with word embedding and pseudo-relevance feedback (PRF). Our results show that PRF-based reranking outperformed word embedding based approach and a strong baseline using inter-wiki link frequency. A hybrid method, Rank Score Combination, achieved the best results. Our analysis also suggests that medical synonyms mined from Wikipedia can increase the coverage of existing synonym resources such as UMLS.

1 Introduction

Biomedical synonym resources have been an important part of biomedical natural language processing (NLP). Synonym resources have been used for a variety of tasks such as query expansion (Aronson and Rindflesch, 1997; Díaz-Galiano et al., 2009), reformulation (Plovnick and Zeng, 2004), and word sense disambiguation (McInnes et al., 2007).

Another important avenue of their use lies in e-portals for clinical notes such as My HealtheVet patient portal, which allows patients to access clinical notes written by their healthcare providers (Nazi et. al., 2013). While many organizations have been embracing these methods of patient-clinician communication, various studies (Lerner et al., 2000; Chapman et al., 2003; Keselman et al., 2007) have shown that patients often have difficulty in comprehending clinical notes.

A patient's ability to comprehend clinical notes is directly related to his/her ability to understand medical jargon (Pyper et al., 2004; Keselman et al., 2007). Subsequently approaches have been developed to replace medical jargon with corresponding lay terms (Kandula et al., 2010; Abrahamsson et al., 2014). Such approaches rely on high quality synonym resource(s). The widely used biomedical knowledge resource, Unified Medical Language System (UMLS) (Humphrey et al., 1998) is a very valuable resource for such purposes. The UMLS incorporates over 100 biomedical terminology resources including Consumer Health Vocabulary (CHV). It also contains definitions for medical terms which can be used to simplify the clinical notes (Ramesh et al., 2013). Even though UMLS is a rich resource with a vast quantity of medical terms, we found that several synonymous or related medical terms that we extracted through Wikipedia, were not present in the UMLS dictionaries. We report this coverage in Section 5.2.

In this paper, we propose a data-driven approach for automatic extraction and ranking of medical synonyms from Wikipedia. Wikipedia is a free-access, free-content collaborative online encyclopedia. Our previous work suggests that about 40% content in Wikipedia contain health related information (Liu et al., 2013). Many studies have shown that Wikipedia contains high quality of biomedical content (Reavley et al., 2012; Devgan et al., 2007; Rajagopalan et al., 2011). For

Proceedings of the Sixth International Workshop on Health Text Mining and Information Analysis (Louhi), pages 142–151,
Lisbon, Portugal, 17 September 2015. ©2015 Association for Computational Linguistics.

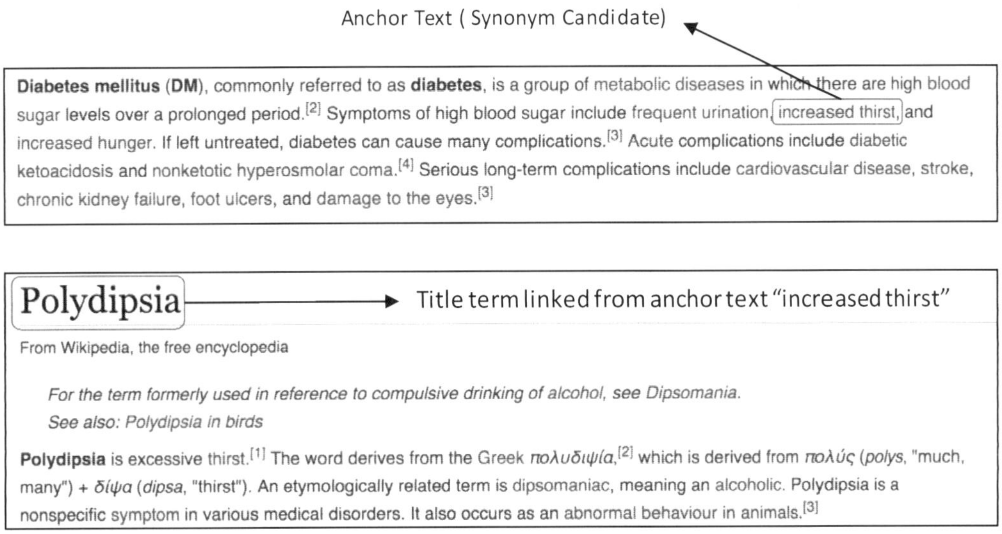

Figure 1: Introductory paragraph for the "Diabetes Mellitus" page in English Wikipedia on top, along with the "Polydipsia" Page below it

example, Devgan et al. (2007) evaluated that Wikipedia contains highly accurate medical articles. They do however, also mention that some articles contain incomplete medical content. Rajagopalan et al. (2011) concluded that Wikipedia has similar accuracy and depth as a professionally edited database. Similarly, Reavley et al. (2012) showed that Wikipedia contains high quality information on mental disorders.

As the result, Wikipedia is being increasingly used by healthcare providers. Specifically, studies show that Wikipedia is widely used by junior physicians (Hughes et al., 2009) and pharmacists (Brokowski & Sheehan, 2009). Additionally Wikipedia is also being widely used by people who are looking for healthcare information. Based on the search engine ranking and page view statistics, Laurent and Vickers (2009) concluded that English Wikipedia is major source of health related information for online users.

Since Wikipedia is written collaboratively by anonymous volunteers, a majority of whom are lay people, its content contains both biomedical jargons and lay terms. This makes Wikipedia a rich resource linking medical jargon with synonymous lay phrases. We leverage this resource by extracting inter-wiki links from Wikipedia to obtain (page title, anchor text) pairs. A typical Wikipedia page includes a title and a description text in which anchor texts are linked (through inter-wiki links) to other Wikipedia pages. As illustrated in Figure 1, one of the anchor texts in the "Diabetes mellitus" Wikipedia page, "increased thirst" is linked to the corresponding page with the title term "polydipsia". We treat the anchor text as a synonym candidate for the title term, which we treat as target concept.

Synonym candidates and their target concepts extracted from inter-wiki links are often synonymous pairs. For example, the anchor texts "frequent urination," "increased thirst," and "increased hunger" are linked to the title pages of "polyuria," "polydipsia," and "polyphagia", respectively. However, sometimes, the synonym candidates and their target concepts are only related but not synonymous. For example "nonketotic hyperosmolar coma" and "kidney failure" are linked to the "hyperosmolar hyperglycemic state" and "chronic kidney disease" respectively.

In addition, as a crowdsourcing resource, Wikipedia has noise. A typical case is where the inter-wiki links are tagged partially. For example, only the "attack" in "heart attack" may be linked to "Myocardial Infarction".

To improve the quality of synonym extraction, we explored several unsupervised methods to rank the synonymous pairs, which utilize distributed word representation (i.e., word embeddings), pseudo relevance feedback (PRF) based re-ranking, and ranking combinations. To our knowledge, this is the first effort that uses word embedding-based ranking and PRF to improve synonym extraction from Wikipedia. We compared our methods with a strong baseline method which uses entity-link frequency.

2 Background

2.1 Related Work

Synonym identification has been active research for two decades. Landauer and Dutnais (1997) used latent semantic analysis to generate 300-dimension word vectors to rank answers of synonym questions in TOEFL. Turney (2001) used search queries to obtain Point-Wise Mutual Information score for two terms to judge whether they are synonyms. Yu et al. (Yu et al., 2002; Yu and Agichtein, 2003) developed rule-based and learning-based methods for extracting author-defined synonyms from text (e.g., using surface cue phrases such as "also called" and parentheses to identify full synonyms and their abbreviations).

Neelakantan and Collins (2015) applied Canonical Correlation Analysis to calculate representation of phrases which were then used for synonym classification. McCrae and Collier (2008) used automatically generated patterns (regular expressions) to mine candidate synonym pairs, which were then classified as synonymous or not based on the occurrence of term pairs in each pattern. Henriksson et al. (2014) created ensembles of semantic spaces, by combining different distributional models and semantic spaces induced from different corpora, for synonym extraction. Blondel et al. (2004) used a central similarity measure in word graphs to calculate similarity between two words. They constructed their graph using a dictionary with the assumption that synonyms were likely to have common words in their definitions and might simultaneously appear in the definitions of many other words. Wang et al. (2015) modified the word2vec algorithm to create a semi-supervised approach that learned from both unlabeled text corpus and UMLS semantic types, groups.

Bøhn and Nørvåg (2010) used redirect pages and inter-wiki links to extract named entities from Wikipedia. They used the frequency of inter-wiki links and other heuristics (e.g., letter capitalization) to rank the synonym candidates. In our work, we use inter-wiki link frequency as our baseline and study the improvements provided by various methods described in section 3.

2.2 Word Representations

Word representations keep the semantic and contextual information of a word in a compact format (e.g., a vector or a tensor). Different methods have been used to obtain compact representations, including clustering based approaches (e.g., Brown Clustering (Brown et al., 1992)), co-occurrence based approaches (Lebret and Collobert, 2014; Pennington et al., 2014), and hierarchical language models (Mnih and Hinton, 2009). Mikolov et al. (2013a, 2013b) showed that using a dense vector representation for words outperforms methods like tf-idf in NLP tasks, e.g., Microsoft Sentence Completion Challenge (Zweig and Burges, 2011). It is expected that words sharing similar semantics or contexts will be close in the projected latent space. In this study, we used the Skip Gram model (Mikolov et al. 2013a) to compute relatedness of synonym pairs extracted from the Wikipedia. Skip Gram models, which belong to distributed word representation (i.e., word embedding) models, are trained through a log-linear classifier that maximizes the prediction accuracy of words within a certain range before and after the current word. We used word vector based similarity methods to rank the synonym candidates because we believe that it has a better semantic representation than the simpler frequency-based approach.

Medical target concepts in Wikipedia are often linked to a variety of synonym candidates; however we found that for several cases, the number of links for each synonym candidate sometimes is very low. For those cases, frequency of inter-wiki links may not be sufficient to accurately determine the ranking of synonym candidates. For example, the target concept "myopathy", is linked to "exertional myopathy", "hereditary myopathy", "muscle disorders", "muscle weakness", "muscular diseases", "polyneuropathy" and "metabolic myopathy", "progressive myopathy" through inter-wiki links. However, each of these inter-wiki links occurs only once. As a consequence, we cannot rank these synonym candidates using their link frequencies. Word embedding approaches do not suffer from this problem and are expected to perform better in such cases.

In addition, word embedding approaches can filter out frequent but partial synonym candidates and provide better ranking. An example of a partial synonym candidate is the "heart attack" example discussed before, where only the word "attack" is tagged as the anchor-text. We expect that such erroneous synonym candidates are rare occurrences and can be filtered out using their link frequency. But, in reality due to erroneous manual tagging, partial anchor-texts (i.e. synonym candidates) sometimes occur more frequently than the true synonyms. For example, "oral cancer" is linked most frequently through "mouth" and "oral" (eight and four times respectively), while a correct paraphrase like "cancer of mouth and tongue" is only linked one time. Word embedding approaches represent semantics better than the frequency-based approach and therefore may be able to identify synonyms and separate them from false positives.

2.3 Pseudo Relevance Feedback

We use pseudo-relevance feedback (PRF) (Attar and Fraenkel, 1977), a widely used method in information retrieval (IR), to obtain better estimates of the representations of target concept in the latent space. PRF is a subtype of a broader class of methods called relevance feedback models (Rocchio, 1971) in IR. Relevance feedback models exploit the idea of using feedbacks (typically from the user) about the relevancy of the results returned for an initial query, to improve or enrich this query. PRF, in particular, does not require user interaction, but instead uses the top-*k* retrieved documents as an automatic feedback. These top-ranked documents are added to the query, and the search runs again with the updated query. We adapted this approach to solve the problem of ranking synonym candidates, which we will introduce in detail in Section 3.3

3 Methods

To improve synonym extraction from Wikipedia inter-wiki links, we explored different unsupervised approaches, including several new methods, for synonym candidate ranking.

3.1 Entity Link Frequency (ELF)

ELF ranks (target concept, synonym candidate) pairs by their Wikipedia inter-wiki link frequency. More specifically, each synonym candidate is ranked by the number of times it has been used as an anchor-text to link to the target concept. Because the inter-wiki links in Wikipedia are created manually, the link frequency associated with each candidate term is a very strong indicator of the viability of that particular synonym candidate. Noisy inter-wiki links (e.g., "arrhythmia" — "other causes" and "heart attack" — "attack") often have low frequencies; while high frequency terms ("polydipsia" — "excessive thirst") are often good synonym candidates. This method was used as the baseline in our experiments.

3.2 Word-Embedding Based Ranking

We use word vectors to estimate the similarity of two words by computing the cosine similarity of their vectors in the embedded space. Many medical terms, however, are phrases with two to five words. This requires methods to combine individual word vectors into phrases. In this work, given two phrases a and b (represented by "$a_1 \dots a_n$" and "$b_1 \dots b_m$" respectively), we estimate their similarity by using the average cosine distance between each pair of words they contain, as defined in Equation 1,

$$ACS_{ab} = \frac{1}{nm}\left(\sum_{i=1}^{n}\sum_{j=1}^{m} < W(a_i), W(b_j) >\right) \qquad (1)$$

where $W(.)$ is the normalized word vector of an individual word. This can be interpreted as computing the cosine similarity of the two phrase vectors, where a phrase vector is estimated by the mean of the normalized word vectors of the individual words contained in that phrase. We call this method Average Cosine Similarity (ACS).

3.3 Re-ranking based on Pseudo Relevance Feedback (PRF)

A limitation of the word embedding method that we use, Skip Gram model, is that it does not disambiguate word senses. In other words, the vector of a word represents multiple senses of this word. As a consequence, synonym candidates with non-relevant senses (e.g., a non-medical sense of the target concept) could be ranked high by word embedding-based ranking method. To alleviate this problem, we leverage on Relevance feedback to disambiguate our term vectors.

As introduced in the previous section, Pseudo Relevance Feedback (Attar and Fraenkel, 1977), is a popular technique in IR, which expands a given query by the top-n documents retrieved for this query. This updated query is then used to retrieve the documents. We adapted PRF for our problem by collecting the top-n synonym candidates obtained by the ELF method. We then calculate the mean vector of these *n* candidate phrases and the target concept. This mean vector is used as the new query. We then re-rank all synonym candidates by their Average Cosine Similarity (ACS) to this new query. When selecting the top-*n* synonym candidates through ELF, if there are multiple candidates with the same ELF scores, we use ACS to break the tie. For example, if the

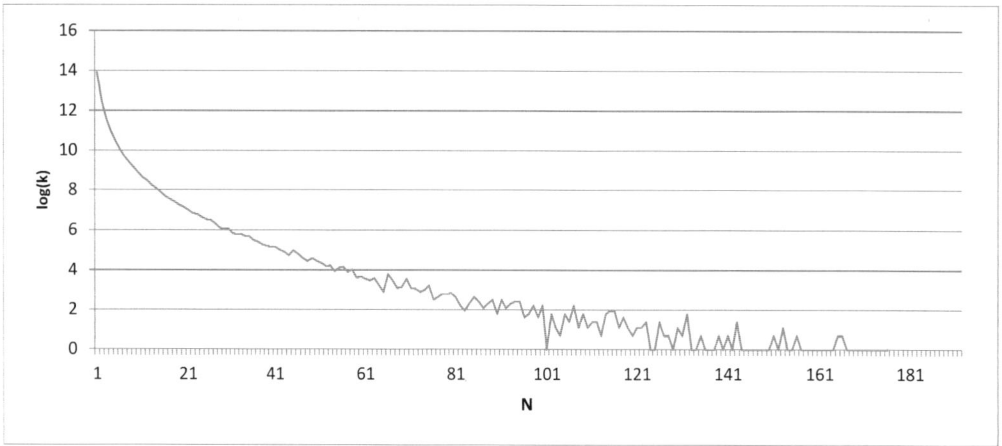

Figure 2: Distribution of number of synonym candidates for Wikipedia Terms. *N* is the number of synonym candidates extracted per title term. Wikipedia Concept. *k* is the number of title terms in Wikipedia that have *N* unique number of synonym candidates

synonym candidates "blood infection" and "bacterial infection" for the target concept "septicemia" have ELF score 1, the candidate with the higher ACS to the concept term will be chosen with a higher priority. In case ACS scores are also tied, we randomly order them.

One major advantage of this method is the use of mean vector of the top-*n* candidates to represent the dominant sense of the target concept. Therefore, it will rank synonym candidates that have senses similar to this dominant sense very highly.

3.4 Ranking Combination

Ensemble ranking is a standard method to combine the strengths of different ranking methods. When we conducted this work, we did not have any annotated data to build a standard ensemble ranker. Instead, we adopted two simple unsupervised methods for ranking combination: (1) Average Ranking (AveR); and (2) Ranking Score Combination (RSC).

AveR ranks the candidates by their mean ranks from ELF, ACS and PRF. RSC ranks the candidates by the sum of their ranking scores from ELF, ACS and PRF. We did not normalize the ELF scores into [0,1] by observing that a large ELF score (i.e., high inter-wiki link frequency) often correctly predicts synonyms, irrespective of its corresponding ACS or PRF scores (which are values that lie between 0 and 1). Our preliminary experiments comparing score combination using normalized vs. raw ELF scores confirmed our choice of using raw ELF scores.

4 Experimental Settings

4.1 Experimental Data

We extracted all the (target concept, synonym candidate) pairs from Wikipedia except the pairs that contain special characters or numbers. In total, we obtained 24M links, with 3.6M unique links for 1.6M distinct concepts.

Figure 2 shows the distribution of the number of synonym candidates extracted for each title term from the Wikipedia. Out of the total 1,659,049 title-terms, 1,457,935 terms have less than three synonym candidates. Our preliminary study suggests that many of these terms are person or location names, which are not of our interest. Therefore, we did not include these terms when creating our gold-standard evaluation dataset and only evaluated our methods on terms with three or more synonym candidates.

We use word2vec software to create the Skip Gram word embeddings. The word embeddings were trained on a combined text corpus of English Wikipedia, Simple English Wikipedia and articles from PubMed Open Access, which contain over 4 billion words in total. The text was lowercased and stripped of all punctuations except comma, apostrophe and period.

We set our word2vec training parameters based on the study of Pyysalo et al. (2013). Specifically, we used 200-dimension vectors with a window size of 6. We used hierarchical soft-max with a subsampling threshold of 0.001 for training.

4.2 Evaluation Dataset

There is no lexical resource suitable for evaluating our task performance. Even UMLS does not cover all the synonyms and related terms we discovered from the Wikipedia. To evaluate our synonym ranking methods, we created a gold standard evaluation dataset from the Wikipedia data we extracted.

Since the goal of this work is to extract synonym candidates for medical terms, we only chose medically relevant concepts for evaluation. We randomly selected 4000 terms from the concepts (title terms) that are present either in the Consumer Health Vocabulary or in the Wikipedia Health Category tree to the depth of 4. An annotator with PhD degree in Biology further selected 1000 relevant medical terms from these 4000 terms.

We built an annotation GUI that presented to the annotators 1000 medical terms and their synonym candidates. Each term and its synonym candidates were shown in a single annotation page. The page order was randomized. The annotation task was to judge whether the synonym candidate was a "Synonym", "Related Term" or "Rejected or Unrelated Term" of the target concept. Two annotators conducted the annotation. Both are premedical school students. So far, 792 unique medical concepts were annotated, out of which 256 were annotated by both. We used these 256 concepts for our evaluation. We also used the entire 792 concepts and their synonyms to calculate the coverage by UMLS.

A synonym candidate is defined as a "Synonym" if it has the exact same meaning as the target concept. It is defined as a "Related Term" if it has a related meaning to the target concept. We accept hypernyms, hyponyms, and words derived from the same root as "Related terms". Additionally we also accept words with high correlations to the target concept, e.g., a very common symptom for a disease. As an example, "high blood sugar" is a related term of "diabetes mellitus". Candidates not in the above-mentioned categories were annotated as "Unrelated or Rejected Terms".

The gold standard of 256 concepts consists of 1507 (title term, synonym candidate) pairs and their corresponding annotations. The linear weighted kappa for the inter-annotator agreement was 0.4762, with the 95% confidence interval ranges from 0.4413 to 0.5111. This kappa value suggests that the annotators have moderate agreement (Viera and Garret, 2005). If we combine related terms and rejected terms into one category, the resulting dataset has a much higher kappa of 0.6250. This contrasts with a low kappa of 0.3929 when related terms are instead combined with synonyms, suggesting that more annotator uncertainty lies in the boundary between related and rejected terms than between related and synonymous terms.

4.3 Evaluation Measure

We use mean average precision (MAP) to evaluate the performances of our ranking methods, because our problem is similar to a typical Information Retrieval tasks. Instead of using a set of relevant and irrelevant documents to evaluate our ranking output, we use a set of synonyms, related terms and rejected terms from our gold-standard annotation for evaluation.

We set two evaluation conditions: (1) combining the synonyms and related terms from the gold-standard annotation to form the set of relevant (positive) instances and treating rejected terms as irrelevant (negative) instances; and (2) using the synonyms from the gold annotation as positive (relevant) instances and treating the related and rejected terms as irrelevant (negative) instances.

By the above definition, condition 1 is a relaxed condition and condition 2 is strict. For both conditions, only terms that were judged by both annotators as relevant (positive) instances are treated as positive.

We compute MAP by Equations (2) and (3).

$$AveP = \sum_{k=1}^{n} P(k)\Delta_r(k) \quad (2)$$

$$MAP = \frac{\sum_{t=1}^{M} AveP(t)}{M} \quad (3)$$

where *AveP* is the average precision of a query (target concept in our case); *k* is the rank of the synonym candidates; *P(k)* is the precision of the ranking at rank *k*; $\Delta_r(k)$ is the increase of recall of the ranking at rank *k* compared with the recall at rank *k-1*; *MAP* is the mean *AveP* of all the target concepts to evaluate on.

5 Results

5.1 Synonym Candidate Ranking

The ranking performances (measured by MAP) of different methods are shown in Table 1.

As we see, under the relaxed condition (Column 1), the word embedding-based ranking method ACS outperforms the frequency based ranking method ELF (Row 2 vs. Row 1); while under the strict condition (Column 2), ACS has slightly lower performance than ELF (Row 2 vs.

Methods	*MAP (Relaxed condition)*	*MAP (strict condition)*
ELF	0.6267	0.2401
ACS	0.6624	0.2383
PRF	0.6859	0.2519
AveR	0.6685	0.2433
RSC	**0.6900**	**0.2745**

Table 1: Mean Average Precision values for Relevance Feedback of 5

Candidates for "septicemia"	*Annotation*	*PRF Ranking*	*Frequency*
bacterial infection	Related	2	1
blood infection	Synonym	6	1
coral poisoning	Rejected	7	1
Septicaemia	Synonym	1	1
Septicaemic	Related	4	1
Septicemic	Related	3	1
septic infection	Synonym	5	1

Table 2: Predictions for "septicemia"

Row 1). This suggests that the word embedding-based ranking method is superior than the frequency based ranking method in identifying semantically related (coherent) terms. However, they themselves may not be sufficient to accurately identify synonyms.

Result analysis suggests that ELF has a high precision at high ranks, especially when the frequency of the candidate term (i.e., the number of times it is linked to the target term in Wikipedia) is high. However, the frequency values for synonym candidates tend to be identical for lower ranked candidates. As a result, it is impossible to determine the order of these candidates using frequency based method such as ELF. Table 2 shows a typical example. For the target concept "septicemia", the frequencies of its candidates are all 1's. In this case, we cannot gain any information from ELF about the ranking of these synonym candidates. The annotated rankings from one of our annotators and the rankings predicted by the PRF method are given on the side. This is a major reason why ELF has lower MAP than ACS and PRF.

PRF performs better than ELF and ACS consistently on both conditions. As introduced in Section 3, in the PRF method, we use the top-*n* (*n*=5 in our experiments) candidate terms returned by the ELF method as feedback terms and use ACS to break the tie (when there are candidates with the same ELF scores). This way, PRF implicitly takes advantages of both ELF and ACS, which explains why it is better than these two methods.

Further analysis of the results suggest that PRF is good at rejecting unrelated terms, but can be confused between synonyms and related words. This is especially true when the related terms are just morphologically different from the original term (see Table 2 for an example).

Table 1 also shows the performance from combining individual ranking methods. As we can see, the performance of the average ranking method using equal weights (AveR, Row 4) falls between the best and the worst individual ranking methods on both conditions. This is not surprising because we did not tune the combination weights. It is likely we can boost the ranking performance by optimizing the combination weights using annotated training data, which will be our future work. Interestingly, the performance from using combined ranking scores (RSC) is almost always higher than all the individual methods with respectable margins on both conditions. This result suggests that augmenting ELF rankings with word similarity based measures and pseudo relevance feedback is a very effective way to improve the quality of synonym candidate ranking. Paired t-test shows that our best performing method RSC is significantly better than the ELF baseline on both conditions (p- value$<$0.001). Other methods are significantly better than ELF on the relaxed condition (p-value $<$0.01) but not on the strict condition.

In our experiments, we set *n*, the number of feedback terms used by PRF, as 5. This value was set heuristically due to the lack of the training data. In a post-experiment analysis, we tested the effects of using different values of *n* (from 1 to 10). Figures 3 and 4 show the results. As we can see, the values of *n* do not affect the ranking results remarkably, especially on the strict condition. In particular, the orders of the performances of different methods remain the same.

5.2 Coverage of Synonym Extraction

To estimate how much the Wikipedia based synonym extraction can contribute to existing synonym resources, we analyzed the coverage of our synonyms in UMLS. So far, we have 5025 unique pairs of medical concepts and their synonym terms, which have been annotated (judged) by at least one annotator. Of the 5025 pairs, 4447 have been annotated as either a synonym or a related term. Of these 4447 terms, only 2621 are covered in UMLS.

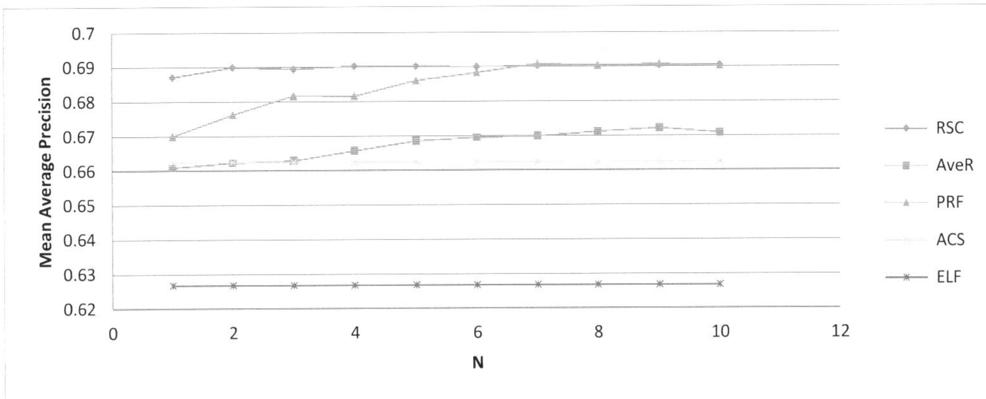

Figure 3: Plot of Mean Average Precision vs N for relaxed condition. N is the number of queries used for Relevance Feedback

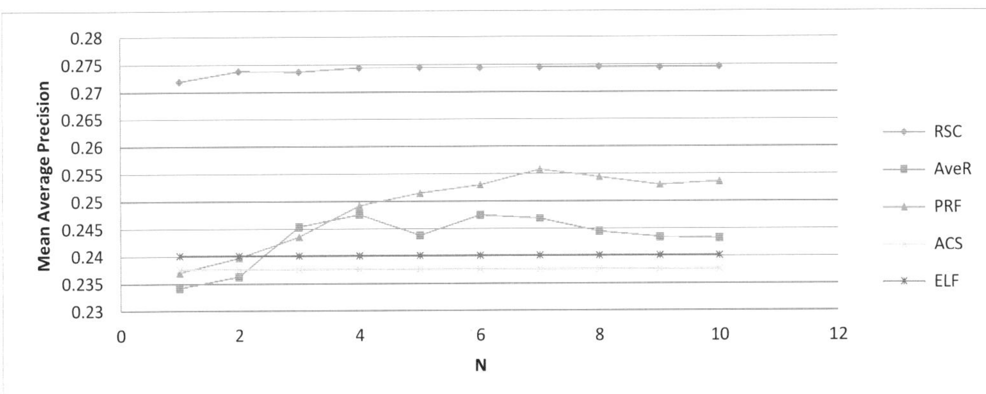

Figure 4: Plot of Mean Average Precision vs N for strict condition. N is the number of queries used for the Relevance Feedback

If we look at only synonyms, 1523 of the 5025 pairs have been annotated by at least one of the annotators as synonyms. Out of the 1523 terms, 429 are not in UMLS.

Clearly Wikipedia is a valuable synonym resource that can be mined to enhance existing lexical resources such as UMLS.

6 Conclusion and Future Work

We have presented novel methods in mining and ranking synonyms from Wikipedia. Our approach is distinguished from previous works in that we utilize word embeddings and pseudo relevance feedback to estimate the semantic and contextual similarities of medical terms and use them as a feature to improve synonym candidate ranking. Our results show that a combination of frequency-based ranking, word embedding based ranking and pseudo relevance feedback achieves the best performance. This suggests that word embedding is a valuable tool in improving synonym extraction from noisy resources like Wikipedia.

We used English Wikipedia for this work. Our approach is general and can be applied to other languages. Its performance is contingent on the size of the Wikipedia and the quality of word embeddings for each specific language. Wikipedia has more than 280 languages, 50 of which have more than hundreds of thousands of articles. The word2vec tool can be trained and used on corpora in any of these languages.

We use the mean of individual word vectors to estimate the phrase vector. In the future, we will explore more advanced algorithms (e.g., Recursive Neural Networks (Socher et al., 2011)) for phrase composition.

The synonym pairs mined and ranked by our methods will be added to a comprehensive synonym resource after manual curation. We will use this resource to simplify medical health records, by substituting complex medical terms with their lay language synonyms.

Acknowledgements

We would like to thank our annotation team (Elaine Freund, Victoria Wang and Shreya Makkapati) for creating the gold-standard evaluation set used in this work.

This work was supported in part by the Award 1I01HX001457 from the United States Department of Veterans Affairs Health Services Research and Development Program Investigator Initiated Research. The contents do not represent the views of the U.S. Department of Veterans Affairs or the United States Government.

References

Abrahamsson, E., Forni, T., Skeppstedt, M., & Kvist, M. (2014). Medical text simplification using synonym replacement: Adapting assessment of word difficulty to a compounding language. *Proceedings of the 3rd Workshop on Predicting and Improving Text Readability for Target Reader Populations (PITR)*, 57–65.

Aronson, A. R., & Rindflesch, T. C. (1997). Query expansion using the UMLS Metathesaurus. *Proc AMIA Annu Fall Symp*, 485–9.

Attar, R., & Fraenkel, A. S. (1977). Local Feedback in Full-Text Retrieval Systems. *J. ACM*, *24*(3), 397–417.

Blondel, V., Gajardo, A., Heymans, M., Senellart, P., & Van Dooren, P. (2004). A measure of similarity between graph vertices. *arXiv:cs/0407061*.

Bohn, C., & Nørvåg, K. (2010). Extracting Named Entities and Synonyms from Wikipedia. In *Advanced Information Networking and Applications* (pp. 1300–1307).

Brokowski, L., & Sheehan, A. H. (2009). Evaluation of pharmacist use and perception of Wikipedia as a drug information resource. *The Annals of Pharmacotherapy*, *43*(11), 1912–1913.

Brown, P. F., Desouza, P. V., Mercer, R. L., Pietra, V. J. D., & Lai, J. C. (1992). Class-based n-gram models of natural language. *Computational Linguistics*, *18*(4), 467–479.

Chapman, K., Abraham, C., Jenkins, V., & Fallowfield, L. (2003). Lay understanding of terms used in cancer consultations. *Psycho-Oncology*, *12*(6), 557–566.

Devgan, L., Powe, N., Blakey, B., & Makary, M. (2007). Wiki-Surgery? Internal validity of Wikipedia as a medical and surgical reference. *Journal of the American College of Surgeons*, *205*(3, Supplement), S76–S77.

Díaz-Galiano, M. C., Martín-Valdivia, M. T., & Ureña-López, L. A. (2009). Query expansion with a medical ontology to improve a multimodal information retrieval system. *Computers in Biology and Medicine*, *39*(4), 396–403.

Henriksson, A., Moen, H., Skeppstedt, M., Daudaravičius, V., & Duneld, M. (2014). Synonym extraction and abbreviation expansion with ensembles of semantic spaces. *Journal of Biomedical Semantics*, *5*(1), 6.

Hughes, B., Joshi, I., Lemonde, H., & Wareham, J. (2009). Junior physician's use of Web 2.0 for information seeking and medical education: A qualitative study. *International Journal of Medical Informatics*, *78*(10), 645–655.

Humphrey, B., Lindberg, D. A. B., Schoolman, H. M., & Barnett, G. O. (1998). The Unified Medical Language System: An Informatics Research Collaboration. *Journal of the American Medical Association*, *5*, 1–11.

Kandula, S., Curtis, D., & Zeng-Treitler, Q. (2010). A Semantic and Syntactic Text Simplification Tool for Health Content. *AMIA Annual Symposium Proceedings*, *2010*, 366–370.

Keselman, A., Tse, T., Crowell, J., Browne, A., Ngo, L., & Zeng, Q. (2007). Assessing consumer health vocabulary familiarity: an exploratory study. *Journal of Medical Internet Research*, *9*(1), e5.

Landauer, T. K., & Dutnais, S. T. (1997). A solution to Plato's problem: The latent semantic analysis theory of acquisition, induction, and representation of knowledge. *PSYCHOLOGICAL REVIEW*, *104*(2), 211–240.

Laurent, M. R., & Vickers, T. J. (2009). Seeking Health Information Online: Does Wikipedia Matter? *Journal of the American Medical Informatics Association*, *16*(4), 471–479.

Lebret, R., & Collobert, R. (2014). Word Embeddings through Hellinger PCA. *Proceedings of the 14th Conference of the European Chapter of the Association for Computational Linguistics*, 482–490.

Lerner, E. B., Jehle, D. V., Janicke, D. M., & Moscati, R. M. (2000). Medical communication: do our patients understand? *The American Journal of Emergency Medicine*, *18*(7), 764–766.

Liu, F., Moosavinasab, S., Agarwal, S., Bennett, A. S., & Yu, H. (2013). Automatically identifying health- and clinical-related content in wikipedia. *Studies in Health Technology and Informatics*, *192*, 637–641.

McCrae, J., & Collier, N. (2008). Synonym set extraction from the biomedical literature by lexical pattern discovery. *BMC Bioinformatics*, *9*(1), 159.

McInnes, B. T., Pedersen, T., & Carlis, J. (2007). Using UMLS Concept Unique Identifiers (CUIs) for Word Sense Disambiguation in the Biomedical Domain. *AMIA Annual Symposium Proceedings*, *2007*, 533–537.

Mikolov, T., Chen, K., Corrado, G., & Dean, J. (2013). Efficient Estimation of Word Representations in Vector Space. *arXiv:1301.3781 [cs]*

Mikolov, T., Sutskever, I., Chen, K., Corrado, G. S., & Dean, J. (2013). Distributed representations of words and phrases and their compositionality. In *Advances in Neural Information Processing Systems* (pp. 3111–3119).

Mnih, A., & Hinton, G. E. (2009). A Scalable Hierarchical Distributed Language Model. In D. Koller, D. Schuurmans, Y. Bengio, & L. Bottou (Eds.), *Advances in Neural Information Processing Systems 21* (pp. 1081–1088). Curran Associates, Inc.

Moen, S. P. F. G. H., & Ananiadou, T. S. S. (n.d.). Distributional Semantics Resources for Biomedical Text Processing.

Nazi, K. M., Hogan, T. P., McInnes, D. K., Woods, S. S., & Graham, G. (2013). Evaluating patient access to Electronic Health Records: results from a survey of veterans. *Medical Care*, *51*(3 Suppl 1), S52–56.

Neelakantan, A., & Collins, M. (2015). Learning Dictionaries for Named Entity Recognition using Minimal Supervision. *arXiv:1504.06650 [cs, Stat]*.

Pennington, J., Socher, R., & Manning, C. D. (2014). Glove: Global vectors for word representation. *Proceedings of the Empiricial Methods in Natural Language Processing (EMNLP 2014)*, *12*.

Plovnick, R. M., & Zeng, Q. T. (2004). Reformulation of consumer health queries with professional terminology: a pilot study. *Journal of Medical Internet Research*, *6*(3), e27.

Polepalli Ramesh, B., Houston, T. K., Brandt, C., Fang, H., & Yu, H. (2013). Improving Patients' Electronic Health Record Comprehension with NoteAid. In *MedInfo* (Vol. 192, pp. 714–718). IOS Press.

Pyper, C., Amery, J., Watson, M., & Crook, C. (2004). Patients' experiences when accessing their on-line electronic patient records in primary care. *The British Journal of General Practice*, *54*(498), 38–43.

Rajagopalan, M. S., Khanna, V. K., Leiter, Y., Stott, M., Showalter, T. N., Dicker, A. P., & Lawrence, Y. R. (2011). Patient-Oriented Cancer Information on the Internet: A Comparison of Wikipedia and a Professionally Maintained Database. *Journal of Oncology Practice*, *7*(5), 319–323.

Reavley, N. J., Mackinnon, A. J., Morgan, A. J., Alvarez-Jimenez, M., Hetrick, S. E., Killackey, E., ... Jorm, A. F. (2012). Quality of information sources about mental disorders: a comparison of Wikipedia with centrally controlled web and printed sources. *Psychological Medicine*, *42*(08), 1753–1762.

Rocchio, J. (1971). Relevance feedback in information retrieval. In *The Smart Retrieval System: Experiments in Automatic Document Processing.* (pp. 313–323).

Socher, R., Huang, E. H., Pennin, J., Manning, C. D., & Ng, A. Y. (2011). Dynamic pooling and unfolding recursive autoencoders for paraphrase detection. In *Advances in Neural Information Processing Systems* (pp. 801–809).

Turney, P. D. (2001). *Mining the Web for Synonyms: PMI-IR Versus LSA on TOEFL.*

Viera, A. J., & Garrett, J. M. (2005). Understanding interobserver agreement: the kappa statistic. *Family Medicine*, *37*(5), 360–363.

Wang, C., Cao, L., & Zhou, B. (2015). Medical Synonym Extraction with Concept Space Models. *arXiv:1506.00528 [cs]*.

Yu, H., & Agichtein, E. (2003). Extracting synonymous gene and protein terms from biological literature. *Bioinformatics (Oxford, England)*, *19 Suppl 1*, i340–349.

Yu, H., Hripcsak, G., & Friedman, C. (2002). Mapping Abbreviations to Full Forms in Biomedical Articles. *Journal of the American Medical Informatics Association*, *9*(3), 262–272.

Zweig, G., & Burges, C. J. C. (2011). *The Microsoft Research Sentence Completion Challenge* (No. MSR-TR-2011-129).

Representing Clinical Notes for Adverse Drug Event Detection

Aron Henriksson
Department of Computer and Systems Sciences
Stockholm University
Sweden
aronhen@dsv.su.se

Abstract

Electronic health records have emerged as a promising source of information for pharmacovigilance. Adverse drug events are, however, known to be heavily underreported, which makes it important to develop capabilities to detect such information automatically in clinical text. While machine learning offers possible solutions, it remains unclear how best to represent clinical notes in a manner conducive to learning high-performing predictive models. Here, 42 representations are explored in an empirical investigation using 27 real, clinical datasets, indicating that combining local and global (distributed) representations of words and named entities yields higher accuracy than using either in isolation. Subsequent analyses highlight the relative importance of various named entity classes for predicting adverse drug events.

1 Introduction

Electronic health records (EHRs) have emerged as a potentially valuable, and complementary, source of information for pharmacovigilance, which, as a result of the limitations of clinical trials – in terms of duration and sample size – needs to be carried out throughout the life-cycle of a drug to inform decisions about sustained use. Adverse drug events (ADEs), defined as undesired harms resulting from the use or misuse of a drug (Nebeker et al., 2004), are the most common iatrogenic injury, being responsible for around 3.7% of hospital admissions worldwide (Howard et al., 2007). The adverse effects of drugs cause suffering in patients and put an economic burden on healthcare – often unnecessarily, as ADEs are in many cases preventable (Hakkarainen et al., 2012).

A challenge for pharmacovigilance is that ADEs are heavily underreported (Hazell and Shakir, 2006), both in so-called spontaneous reporting systems, whereby reports of ADE cases are submitted voluntarily by patients and clinicians, and in EHRs, in which ADEs can be encoded by a set of diagnosis codes. To address the problem of underreporting, systems that can automatically detect ADEs in EHRs are potentially valuable, and much research has been conducted to that end (Harpaz et al., 2012). While many efforts have aimed at using machine learning for detecting ADEs on the basis of structured EHR data (Chazard et al., 2011; Zhao et al., 2014a; Zhao et al., 2014b; Zhao et al., 2015), attempts have also been made to exploit the more unstructured data in the form of clinical notes (Eriksson et al., 2013; LePendu et al., 2013). These have either relied on manually constructed rules and extensive dictionaries or on applying disproportionality methods[1] to counts of terms extracted from clinical notes. In a recent study (Henriksson et al., 2015a), information pertaining to ADEs – including named entities such as drugs and medical problems, as well as relations between them, i.e., whether one exists and whether it expresses, e.g., an indication or an ADE – were detected in clinical notes using machine learning; this approach, however, relies on the availability of data that has been manually labeled outside the clinical setting. There have also been efforts to combine information from the structured and unstructured sections of EHRs for ADE detection (Harpaz et al., 2010; Eriksson et al., 2014). In one of these (Henriksson et al., 2015b), heterogeneous types of clinical data, including free-text notes,

[1] Disproportionality methods describe to what extent the co-occurrence frequency of two events deviates from what is expected (Suling and Pigeot, 2012).

Proceedings of the Sixth International Workshop on Health Text Mining and Information Analysis (Louhi), pages 152–158, Lisbon, Portugal, 17 September 2015.

were represented using distributional semantics, the use of which is also investigated in this study. In the previous study, however, many possible alternative ways of representing clinical notes were left unexplored. A more in-depth investigation is conducted in the present study, focusing on the representation of clinical notes for ADE detection.

In this study, ADE detection using clinical notes is approached as a binary classification task, in which the presence or absence of a particular ADE in a healthcare episode is to be determined; for this purpose, diagnosis codes assigned in the clinical setting are used as class labels. This raises the question of how best to represent clinical notes. There are certainly challenges involved in applying machine learning to high-dimensional and sparse data, which, as a result of prevalent misspellings and creative shorthand, clinical notes are a prime example of. These challenges will be considered when exploring possible representations of clinical notes.

2 Materials and Methods

This study explores 42 different ways of representing clinical notes and evaluates their effectiveness, in terms of classification accuracy, on the task of detecting the presence of an ADE in a healthcare episode. The use of both local and global (distributed) representations of words and named entities, as well as their combination, is investigated in an experiment using 27 ADE datasets, followed by a number of further analyses. Local representations are ones that do not incorporate any prior (semantic) knowledge of the similarity of token types, while global representations do, in this case by applying models of distributional semantics to a much larger corpus, resulting in word embeddings that are then exploited in the ADE detection task. While local representations are commonly employed for document classification, the use of global, distributed representations has been less thoroughly investigated, with a few exceptions (Sahlgren and Cöster, 2004; Henriksson et al., 2015b). Here, various types of local and global representations are compared and combined in an exploratory fashion.

2.1 Data Source

The 27 datasets were extracted from a Swedish EHR database (Dalianis et al., 2012), which contains health records over a two-year period from Karolinska University Hospital[2]. The learning task is to detect healthcare episodes that involve a certain ADE, i.e., in which an ADE-specific ICD-10 diagnosis code has been assigned. A healthcare episode is here defined based on the time interval between recorded activities for a patient, delimited by at least three days of inactivity. Each of the 27 datasets thus consists of healthcare episodes, where the positive examples have been assigned an ADE-related diagnosis code, and the negative examples are an equal number of randomly selected healthcare episodes in which that same code has not been assigned. The ADE-related diagnoses were selected on the basis of having been classified as indicating ADEs in a previous study (Stausberg and Hasford, 2011) and being sufficiently frequent (> 10) in the database. The datasets are described in Table 1. In addition to the labeled datasets, the entire two years of data is used for obtaining global, distributed representations of words. The notes, containing approximately 3M unique types (700M tokens), are preprocessed by using Stagger (Östling, 2013) for tokenization and lemmatization of Swedish text.

2.2 Data Representations

$14 \times 3 = 27$ representations of clinical notes are explored. Each of the fourteen representations of words and/or named entities are weighted in three different ways. The local representations include the commonly employed unigrams, bigrams and trigrams, as well as their combination. In addition, a named entity recognition (NER) model trained on Swedish clinical text (Henriksson et al., 2015a) is applied to the healthcare episodes to extract mentions of the following named entity types: *Finding*, *Disorder*, *Drug*, *Body Part* and *ADE Cue*[3]. Local representations of identified named entities, without specifying type (denoted Terms), as well as a combination of unigrams and terms,

[2]This research has been approved by the Regional Ethical Review Board in Stockholm (Etikprövningsnämnden i Stockholm), permission number 2012/834-31/5.

[3]An ADE Cue corresponds to an expression that indicates the presence of an ADE without revealing its precise manifestation, e.g., *side effect* or *drug-induced*.

Dataset	Episodes	Unigrams		Bigrams		Trigrams		Terms	
		Types	Tokens	Types	Tokens	Types	Tokens	Types	Tokens
D64.2	416	58,455	2,642,271	432,235	2,380,942	895,864	2,121,106	18,204	187,885
E27.3	34	10,990	142,365	55,922	127,740	78,222	113,136	2,729	11,907
F11.0	76	13,906	285,629	73,688	257,154	107,216	228,707	3,532	21,448
F11.2	308	35,340	1,118,138	234,855	1,005,677	422,609	893,405	10,620	83,574
F13.0	120	16,759	264,555	93,847	238,388	144,340	212,272	4,204	18,262
F13.2	76	14,226	262,901	73,413	237,228	106,560	211,607	3,546	19,140
F15.0	32	6,498	48,919	25,674	43,859	31,226	38,818	1,362	3,755
F15.1	46	10,849	136,093	51,897	123,224	72,081	110,391	2,438	9,776
F15.2	256	30,098	812,312	193,693	729,918	340,491	647,691	8,896	61,548
F19.0	122	18,257	341,225	100,890	307,757	152,834	274,353	4,671	23,638
F19.1	74	14,279	216,583	75,268	194,519	109,675	172,500	3,457	16,504
F19.2	288	34,925	992,236	229,330	891,735	404,862	791,349	10,645	76,797
F19.9	68	14,938	221,480	78,506	198,688	112,658	175,942	3,571	16,205
G24.0	28	11,293	125,342	57,711	112,935	81,196	100,542	2,897	9,454
G62.0	20	5,121	44,776	19,564	40,609	23,359	36,449	1,027	3,350
I95.2	70	13,321	179,622	69,127	161,505	99,732	143,442	3,336	14,651
L27.0	274	41,669	1,394,815	281,601	1,255,535	513,539	1,116,442	12,908	109,896
L27.1	78	15,495	291,266	84,668	261,351	125,632	231,491	4,118	23,824
N14.1	28	10,383	101,969	49,864	92,098	67,286	82,247	2,585	7,979
O35.5	128	11,810	145,344	57,172	131,075	79,992	116,826	2,745	10,313
T59.9	40	6,355	57,773	26,831	51,763	34,017	45,766	1,474	4,492
T78.2	102	15,272	236,489	80,533	212,753	118,274	189,062	3,845	19,461
T78.3	266	26,716	503,385	161,366	451,831	265,928	400,439	7,912	43,832
T78.4	1520	56,244	1,950,200	396,818	1,752,142	783,547	1,555,017	18,415	167,620
T80.8	732	48,299	2,053,152	349,030	1,845,434	698,814	1,638,072	16,247	169,391
T88.6	96	17,453	280,652	96,546	252,705	145,766	224,818	4,714	23,191
T88.7	564	51,922	1,422,450	357,484	1,600,899	680,750	1,422,450	16,738	138,899

Table 1: Description of the 27 ADE datasets used in the study

are explored.

In addition to the local representations, the use of global, distributed representations of words and terms is explored. Word embeddings are obtained using a recently introduced model of distributional semantics – see (Cohen and Widdows, 2009) for an overview – based on shallow neural networks with a single hidden layer: the skip-gram model (Mikolov et al., 2013) as implemented in word2vec. It was chosen for its ability to produce high-quality vector representations of words, outperforming traditional context-counting based methods on a range of tasks (Baroni et al., 2014). The algorithm obtains vector representations of the words in the training set by learning to predict nearby context words of each target word; the learned weights within the neural network are then used as vector representations. In a basic configuration, a symmetric context window size of 10 and a dimensionality of 200 is employed[4]. Distributed representations of clinical notes are obtained by simply summing the vectors corresponding to the constituent token types; when representing notes by terms, the words that make up multiword terms are likewise summed. As it has been shown that improved performance can be obtained by combining various word representations (Henriksson et al., 2014), we also explore the use of distributed ensembles created by employing a number of different context window sizes: 6, 8, 10, 12, 14. The representations of healthcare episodes are then obtained by fusing the features from each distributional semantic space. The intuition behind this is that they will capture different aspects of the data. Both single distributed representations and ensembles thereof are used to model healthcare episodes as a combination of unigrams and terms.

Finally, combinations of local and global representations are explored: (1) combining local and global representations of unigrams and terms from a single semantic space, and (2) combining them from multiple semantic spaces. In all representations, the lowercase lemma of the tokens is used. The three weighting strategies are: (1) binary, (2) term frequency (TF), and (3) term frequency-inverse document frequency (TF-IDF). The binary representation corresponds to the so-called one-of-K or one-hot encoding,

[4]10 is the "recommended" context window size for the skip-gram model; employing a higher dimensionality generally, but not necessarily, leads to better representations (Mikolov, 2015).

indicating the presence or absence of a feature; TF corresponds to the bag-of-words representations; finally, TF-IDF is the product of TF in a particular document and the term's IDF. It thus gives less weight to common terms with little discriminative value.

2.3 Experimental Setup

The main experiment involves a comparison of the 42 representations and their impact on classification accuracy. Here, the random forest algorithm (Breiman, 2001) is used due to its reputation of achieving high accuracy, its ability to handle high-dimensional data, as well as the possibility of obtaining estimates of variable importance. The algorithm constructs an ensemble of decision trees, which together vote for what class label to assign to an example. Each tree in the forest is built from a bootstrap replicate of the original instances, while a subset of all features is sampled at each node when building the tree. This procedure is intended to increase diversity among the trees. When the number of trees in the forest increases, the probability that a majority of trees makes an error decreases, given that the trees perform better than random and that the errors are made independently. Although this can only be guaranteed in theory, the algorithm has often been shown in practice to result in state-of-the-art predictive performance. In this study, we use random forest with 500 trees, while $\sqrt{n}$ of all available n features are inspected at each node.

Using the terms representation, a follow-up analysis is conducted to gain insight into which (types of) terms are most useful in the classification task. Variable importance can be estimated in different ways (Breiman, 2001). Here, Gini importance is used as the variable importance metric, where high Gini importance means that a feature plays a greater role in splitting the data into the defined classes. A Gini importance of zero indicates that a feature is considered useless or is never selected to build any tree. We inspect the twenty most important features, averaged over datasets, but we also calculate the average rank of terms of various lengths and named entity classes to understand which types of terms are more informative. Finally, the frequency of various named entity types across the two classes is analyzed in an attempt to identify potentially important differences.

Models are built and evaluated using ten iterations of stratified 10-fold cross validation. For testing the statistical significance of observed differences between the various representations, the Friedman test, as suggested in (Garcia and Herrera, 2008), is employed, where the null hypothesis is that the methods perform equally well.

3 Results

The accuracy scores, averaged over the 27 datasets, produced with the various data representations are shown in Table 2. A Friedman test rejects the null hypothesis that the various representations perform equally well ($p < 0.0001$). Of the three weighting strategies, the binary strategy perfroms almost invariably better than the TF and TF-IDF strategies. When comparing the ngram representations, unigrams perform considerably better than bigrams and trigrams, while their combination is plausibly negatively affected by the latter two. Using only extracted terms performs slightly better than using all unigrams or a combination of unigrams and terms, albeit the differences are small. The global, distributed representations only outperform the local representations when multiple semantic spaces are used in an ensemble. Moreover, all ensembles outperform their single-model counterparts. The best predictive performance is obtained when combining local and global representations – in a semantic space ensemble – of unigrams and terms, yielding an accuracy of 83.89%.

The twenty most important term features are listed in Table 3. All of these are names of drugs, findings and disorders. Some of the drugs are known to cause ADEs, while others are used for treating ADEs. Many of the highly-ranked terms appear only in a single or a handful of datasets; additional highly-ranked terms that appear in all 27 datasets – and conceivably important for detecting ADEs generally – include *smärta* (Eng: pain), *trött* (Eng: tired), *feber* (Eng: fever) and *utslag* (Eng: rash). Named entity mentions of type *ADE Cue* were ranked somewhat lower (out of ~78k): *reaktion* (Eng: reaction) – 53, *biverkan* (Eng: side effect) – 332, *läkemedelsbiverkan* (Eng: drug reaction) – 855 and *läkemedelsutlöst* (Eng: drug-induced) – 19602. When inspecting

	Binary	TF	TF-IDF
Unigrams	83.05	81.70	81.72
Bigrams	76.65	75.98	75.67
Trigrams	68.13	66.93	67.02
Ngrams (Unigrams + Bigrams + Trigrams)	79.47	78.33	78.43
Terms	83.12	81.47	81.59
Unigrams + Terms	83.09	81.81	81.76
Distributed (Unigrams)	81.13	79.59	78.16
Distributed (Terms)	82.82	82.99	75.12
Distributed Ensemble (Unigrams)	82.23	81.53	79.30
Distributed Ensemble (Terms)	83.51	82.82	75.71
Distributed (Unigrams + Terms)	82.04	80.63	76.84
Distributed Ensemble (Unigrams + Terms)	83.71	82.93	80.78
Unigrams + Terms + Distributed (Unigrams + Terms)	83.31	82.30	82.32
Unigrams + Terms + Distributed Ensemble (Unigrams + Terms)	**83.89**	82.72	82.96

Table 2: Average accuracy (%) over 27 ADE datasets with different representations of clinical notes

the average rank of terms of varying length, unigrams were ranked the highest, followed by bigrams, trigrams and ngrams with $n > 3$. Calculating the average rank of terms of various named entity types revealed that *ADE Cue* was ranked the highest, followed by *Disorder*, *Body Part*, *Drug* and *Finding*.

Rank	Term (Swedish)	Term (English)	NE Type	Support
1	missbruk	addiction	Finding	23
2	bev-fl-iri	bev-fl-iri	Drug	2
3	amfetamin	amphetamine	Drug	20
4	cyanokit	cyanokit	Drug	1
5	läkemedels-utlöst dystoni	drug-induced dystonia	Disorder	1
6	betapred	betapred	Drug	27
7	intox	intoxication	Disorder	22
8	akut dystoni	acute dystonia	Disorder	3
9	hepatit c	hepatitis c	Disorder	27
10	allergisk reaktion	allergic reaction	Disorder	25
11	tavegyl	tavegyl	Drug	25
12	syrgas	oxygen	Drug	27
13	amfetamin-missbruk	amphetamine abuse	Disorder	23
14	mätbar sjukdom	measurable disease	Disorder	1
15	stesosolid	stesosolid	Drug	26
16	svullnad	swelling	Finding	27
17	kontrahera	contract	Finding	1
18	bltr vara stabil	blood pressure be stable	Finding	1
19	klåda	itching	Finding	27
20	hjärtmuskel-inflamation	myocarditis	Disorder	1

Table 3: Variable importance of terms

A means of studying potential differences between the two classes is simply to count the number of terms in the healthcare episode according to their class label. The result of this is shown in Table 4. The number of terms per healthcare episode is considerably higher for the ADE class; however, this can partly be explained by differences in average document length: 3575 tokens for positive episodes and 2098 for negative episodes. A fairer comparison is, then, to calculate the number of tokens per encountered term. This comparison reveals that the numbers of *Drug*, *ADE Cue*, *Body Part* and *Finding* mentions are lower for the ADE class, especially the first two, which means that they are more frequent.

	ADE		Not ADE	
NE Type	Term / Episode	Tokens / Term	Term / Episode	Tokens / Term
Disorder	34.65	103.19	21.57	97.28
Finding	124.14	28.80	68.07	30.83
Drug	74.68	47.87	39.68	52.89
Body Part	49.27	72.57	27.58	76.08
ADE Cue	1.94	1839.01	0.86	2432.98

Table 4: The distribution of terms over classes

4 Discussion

This study explored the use of 42 different representations of clinical notes from healthcare episodes for the automatic detection of adverse drug events. It was shown that combining local and global, distributed representations yielded the highest predictive performance. While the use of a simple unigram model worked well, performance quickly deteriorated as larger ngrams were used, most probably as a result of the ensuing sparsity. Interestingly, using only extracted terms outperformed the use of all unigrams, with the added benefit that the former is much lower-dimensional and thus computationally preferable. Even lower-dimensional – and denser – are the distributed representations: in this case 200 with a single semantic space and 200 × 5 with the semantic space ensemble. A distinct advantage of distributed representations is their scalability, as the dimensionality does not grow with the size of the vocabulary, allowing more information to be exploited effectively, as demonstrated by the distributed ensemble of unigrams and terms. The best results were, however, obtained when combining local and ensembles of global, distributed representations. While the difference to using a simple unigrams model is not very large, it is interesting to note the bigger difference to using the commonly employed bag-of-words representation. The advantage of using a binary representation over TF or TF-IDF weighting was also somewhat surprising but can perhaps be attributed to the noisy nature of clinical text.

An advantage of using the terms representation is that, in comparison to the other representations – in particular the distributed ones – it lends itself to some degree of interpretability. While random forest belongs to a family of opaque models, inspection of variable importance provides some insight. It was not surprising that *ADE Cue* terms were, on average, ranked the highest, although somewhat more so that *Body Part* terms were ranked higher than *Drug* and *Finding* terms. When inspecting the distribution of terms over classes, however, it was confirmed that *Drug* and *ADE Cue* terms were common in ADE episodes than in non-ADE episodes, which seems intuitive. For future work, it would be interesting to study whether enriching the representation with factuality – including negation and uncertainty – and temporality would be lead to improved predictive performance.

Acknowledgments

This work was supported by the project High-Performance Data Mining for Drug Effect Detection at Stockholm University, funded by the Swedish Foundation for Strategic Research under grant IIS11-0053.

References

Marco Baroni, Georgiana Dinu, and Germán Kruszewski. 2014. Don't count, predict! a systematic comparison of context-counting vs. context-predicting semantic vectors. In *Proceedings of the 52nd Annual Meeting of the Association for Computational Linguistics*, volume 1, pages 238–247.

Leo Breiman. 2001. Random forests. *Machine Learning*, 45(1):5–32.

Emmanuel Chazard, Gregoire Ficheur, Stephanie Bernonville, Michel Luyckx, and Regis Beuscart. 2011. Data mining to generate adverse drug events detection rules. *IEEE Transactions on Information Technology in Biomedicine*, 15(6):823–830.

Trevor Cohen and Dominic Widdows. 2009. Empirical distributional semantics: Methods and biomedical applications. *Journal of Biomedical Informatics*, 42(2):390 – 405.

Hercules Dalianis, Martin Hassel, Aron Henriksson, and Maria Skeppstedt. 2012. Stockholm EPR Corpus: A Clinical Database Used to Improve Health Care. In *Proceedings of the Swedish Language Technology Conference (SLTC)*.

Robert Eriksson, Peter Bjødstrup Jensen, Sune Frankild, Lars Juhl Jensen, and Søren Brunak. 2013. Dictionary construction and identification of possible adverse drug events in danish clinical narrative text. *Journal of the American Medical Informatics Association*, 20(5):947–953.

Robert Eriksson, Thomas Werge, Lars Juhl Jensen, and Søren Brunak. 2014. Dose-specific adverse drug reaction identification in electronic patient records: temporal data mining in an inpatient psychiatric population. *Drug safety*, 37(4):237–247.

Salvador Garcia and Francisco Herrera. 2008. An Extension on" Statistical Comparisons of Classifiers over Multiple Data Sets" for all Pairwise Comparisons. *Journal of Machine Learning Research*, 9(12).

Katja M Hakkarainen, Khadidja Hedna, Max Petzold, and Staffan Hägg. 2012. Percentage of patients with preventable adverse drug reactions and preventability of adverse drug reactions–a meta-analysis. *PloS one*, 7(3):e33236.

Rave Harpaz, Krystl Haerian, Herbert S Chase, and Carol Friedman. 2010. Mining electronic health records for adverse drug effects using regression based methods. In *Proceedings of the 1st ACM International Health Informatics Symposium*, pages 100–107. ACM.

Rave Harpaz, William DuMouchel, Nigam H Shah, David Madigan, Patrick Ryan, and Carol Friedman. 2012. Novel data-mining methodologies for adverse drug event discovery and analysis. *Clinical Pharmacology & Therapeutics*, 91(6):1010–1021.

Lorna Hazell and Saad AW Shakir. 2006. Under-reporting of adverse drug reactions. *Drug Safety*, 29(5):385–396.

Aron Henriksson, Hans Moen, Maria Skeppstedt, Vidas Daudaravicius, and Martin Duneld. 2014. Synonym extraction and abbreviation expansion with ensembles of semantic spaces. *Journal of Biomedical Semantics*, 5:6:1–25.

Aron Henriksson, Maria Kvist, Hercules Dalianis, and Martin Duneld. 2015a. Identifying Adverse Drug Event Information in Clinical Notes with Distributional Semantic Representations of Context. *Journal of Biomedical Informatics*, in press.

Aron Henriksson, Jing Zhao, Henrik Boström, and Hercules Dalianis. 2015b. Modeling heterogeneous clinical sequence data in semantic space for adverse drug event detection. In *Proceedings of the International Conference on Data Science and Advanced Analytics (DSAA)*. IEEE.

RL Howard, AJ Avery, S Slavenburg, S Royal, G Pipe, P Lucassen, and M Pirmohamed. 2007. Which drugs cause preventable admissions to hospital? a systematic review. *British Journal of Clinical Pharmacology*, 63(2):136–147.

Paea LePendu, Srinivasan V Iyer, Anna Bauer-Mehren, Rave Harpaz, Jonathan M Mortensen, Tanya Podchiyska, Todd A Ferris, and Nigam H Shah. 2013. Pharmacovigilance using clinical notes. *Clinical pharmacology & therapeutics*, 93(6):547–555.

Tomas Mikolov, Kai Chen, Greg Corrado, and Jeffrey Dean. 2013. Efficient estimation of word representations in vector space. *arXiv preprint arXiv:1301.3781*.

Tomas Mikolov. 2015. word2vec: Tool for computing continuous distributed representations of words. `https://code.google.com/p/word2vec/`. Accessed: 2015-08-11.

Jonathan R Nebeker, Paul Barach, and Matthew H Samore. 2004. Clarifying adverse drug events: a clinician's guide to terminology, documentation, and reporting. *Annals of internal medicine*, 140(10):795–801.

Robert Östling. 2013. Stagger: An open-source part of speech tagger for swedish. *Northern European Journal of Language Technology (NEJLT)*, 3:1–18.

Magnus Sahlgren and Rickard Cöster. 2004. Using bag-of-concepts to improve the performance of support vector machines in text categorization. In *Proceedings of the 20th international conference on Computational Linguistics*, page 487. Association for Computational Linguistics.

Jürgen Stausberg and Joerg Hasford. 2011. Drug-related admissions and hospital-acquired adverse drug events in germany: a longitudinal analysis from 2003 to 2007 of icd-10-coded routine data. *BMC Health Services Research*, 11(1):134.

Marc Suling and Iris Pigeot. 2012. Signal detection and monitoring based on longitudinal healthcare data. *Pharmaceutics*, 4(4):607–640.

Jing Zhao, Aron Henriksson, Lars Asker, and Henrik Boström. 2014a. Detecting adverse drug events with multiple representations of clinical measurements. In *International Conference on Bioinformatics and Biomedicine (BIBM)*, pages 536–543. IEEE.

Jing Zhao, Aron Henriksson, and Henrik Boström. 2014b. Detecting adverse drug events using concept hierarchies of clinical codes. In *International Conference on Healthcare Informatics (ICHI)*, pages 285–293. IEEE.

Jing Zhao, Aron Henriksson, and Henrik Boström. 2015. Cascading adverse drug event detection in electronic health records. In *Proceedings of the International Conference on Data Science and Advanced Analytics (DSAA)*. IEEE.

Author Index

Association for Computational Linguistics
209 N. Eighth Street
Stroudsburg, Pennsylvania 18360

ISBN 978-1-5108-1191-1

CURRAN ASSOCIATES INC.
proceedings.com